Free Exam Tips Videos/DVD

We have created a set of videos to better prepare you for your exam. We would like to give you access to these **videos** to show you our appreciation for choosing Exampedia. **They cover proven strategies that will teach you how to prepare for your exam and feel confident on test day.**

To receive your free videos, email us your thoughts, good or bad, about this book. Your feedback will help us improve our guides and better serve customers in the future.

Here are the steps:

    1. Email **freevideos@exampedia.org**

    2. Put "**Exam Tips**" in the subject line

Add the following information in the body of the email:

    3. **Book Title:** The title of this book.

    4. **Rating on a Scale of 1–5:** With 5 being the best, tell us what you would rate this book.

    5. **Feedback:** Give us some details about what you liked or didn't like.

Thanks again!

PSAT Prep 2021 and 2022
Study Guide with Practice Test Questions for the NMSQT Pre SAT College Board Exam
[Book Includes Detailed Answer Explanations]

Andrew Smullen

Copyright © 2021 by Exampedia

ISBN 13: 9781637759189
ISBN 10: 1637759185

All rights reserved. No part of this publication may be reproduced, distributed, or transmitted in any form or by any means, including photocopying, recording, or other electronic or mechanical methods, without the prior written permission of the publisher, except in the case of brief quotations embodied in critical reviews and certain other noncommercial uses permitted by copyright law.

Written and edited by Exampedia.

Exampedia is not an affiliate or partner of any official testing organization. Exampedia is a publisher and distributor of unofficial educational products. All exam and organization names are trademarks of their respective owners. Content in this book exists for utilitarian purposes only and does not necessarily reflect the point of view of Exampedia.

For bulk discounts and any other inquiries, contact: info@exampedia.org

Table of Contents

Preview 1

Test Tips 3

FREE Videos/DVD OFFER 5

Introduction 6

Reading 7

   Command of Evidence 8

   Words in Context 16

   Analysis in History/Social Studies and in Science 20

   Practice Quiz 26

   Answer Explanations 28

Writing and Language 29

   Command of Evidence 30

   Words in Context 31

   Expression of Ideas 38

   Standard English Conventions 41

   Analysis in History/Social Studies and in Science 57

   Practice Quiz 59

   Answer Explanations 61

Math 62

   Heart of Algebra 62

   Problem Solving and Data Analysis 70

   Passport to Advanced Math 84

   Practice Quiz 100

   Answer Explanations 102

Practice Test 103

Reading 103

Writing and Language 119

Math 129

Answer Explanations **142**

Reading 142

Writing and Language 148

Math 153

Preview

Congratulations! You've decided to study this guide in order to become one step closer to your educational or professional goal. This guide is designed to follow your specific test outline or educational format so that you will have the best chance at acing the test or becoming an expert in your field. Equipped with test tips, an introduction to the exam material, and content sections, we will go through each portion of the guide so that you can get a preview before you dive in.

The test tips are actionable items that assist you in preparing for the exam as well as suggestions to utilize during the exam. Studying the outline of your exam, reading thoroughly, looking for test-writing tricks, and keeping calm and collected are the basics needed in having a positive test experience. Having adequate knowledge of not just what's on a test, but *how* to take a test will set you up for success where other test takers may only be minimally prepared.

Knowing your exam content is only one of the many tools you need in order to take a test. Where will you take your test? How will it be scored? Should you still answer a question you don't know? How long will the test take? These questions and more will be answered in the Introduction portion of this guide, which goes over the exam you are studying for, who created it, what percentage of students take it and pass, and how scoring works. You will learn about your test beyond its content so that you can be fully prepared with as little distractions as possible on the day of the test.

This guide goes over the content knowledge needed to take your exam, usually in the order it will be presented on the test. Additionally, practice questions are given at the end of each guide that attempt to mimic your actual exam, along with their answer explanations. It's important that we deliver to you detailed and adequate answer explanations so that if you receive an incorrect answer, you will know why. Remember that you will make mistakes with your test questions, on subtle language or on content knowledge, but practicing taking the exam will help you read the questions more carefully and learn your content like an expert. After taking the practice exam, we hope you will be a well-prepared test taker!

In addition to the test tips below, we want to remind you of a few important things to fully prepare yourself for test-taking day.

- Avoid cramming the night before
- Get a good night's sleep and let your brain fully recover for the test ahead
- Eat a balanced breakfast in the morning
- Arrive at the test center half an hour before test time

During the test, follow these instructions to remain a calm and prepared test taker:

- Pace yourself; don't spend too much or too little time on a single question
- Use all the allotted time you have; go back and check to see that you've answered all relevant questions
- A positive mindset is key; affirmations such as "I can do this" or "I am equipped to pass this exam" will empower you to persevere throughout the day
- Remember to breathe if you find yourself holding your breath; adequate oxygen will help to clear your mind of panic and doubt

Listen—you've got this! Your decision to take time with this study guide is the first step in becoming a confident, successful test taker. Congratulations again; we wish you the best in your educational and professional journey.

If you have any questions or concerns, please feel free to contact us at:

    info@exampedia.org

Sincerely,
Exampedia Test Prep

Test Tips

1. Be Familiar with the Test

Besides being knowledgeable in your exam's content areas, the next best thing is being knowledgeable about the exam itself. Being familiar with the test means that from the time you walk into the testing center to the time you receive the test scores, you have an expectation of how things will be done. Read all about the test in our introduction section to know how much time you have to take it, what sections are in it, and what type of problems you will encounter. Be aware of how the test is scored so that you know whether to take a guess or skip a question. It's important to go into the test knowing exactly what to expect so that you can be a confident test taker.

2. Read the Directions Carefully

Before you begin the exam, read the directions carefully. These will tell you how to answer the questions. For example, are the directions asking you to choose ONE answer only, or AT LEAST TWO answers? Is the word bank showing you ANTONYMS or SYNONYMS, and which should you choose? The directions will tell you everything you need to know about answering the question correctly.

3. Read the Whole Question

It's tempting to look at the answers right away. However, if possible, cover the answers so that full focus can be given to the question first. Read every single word out loud in your head; test writers will sometime insert negatives ("which of the following is *not*..." "All of the following *except*..."), and they can be easy to miss. Reading every single word of the question with intention will help you eliminate choices that might be designed to trick you.

4. Read Every Answer Choice

There are several strategies that go into choosing a correct answer. First, read the answers without any bias; sometimes the test writers will play on test takers' bias to trip them up. A helpful tip for Reading Comprehension passages is to choose the answer choice that is true *in the world of the passage*. If the passage says the sky is purple, don't choose the answer that says the sky is blue. Remember that the passage is where you will find answers. Additionally, eliminate answers you know for a fact are incorrect. With the remaining answers, don't automatically jump to the first one you think is correct. Read all of them, and then choose. Sometimes, one answer may be *more correct* than another one. Read every word of the choices to make sure you've got the best one.

5. Look for Subtle Negatives

Sometimes test writers will insert subtle negatives into the question with words such as *not, except,* or *never*. Subtly reversing the meaning of questions examines the test taker's ability to follow directions and read thoroughly. Test takers who read each word thoroughly will not miss these subtle negatives. Look for these words in order to circle or underline them.

6. Look for Key Words

If you're taking an exam on paper, circle or underline the key words that will help you to answer the question. If you are on a computer and have scratch paper, jot down key words from the question. For Social Studies or Reading Comprehension questions, you can look for words like "main idea," "theme," "organization," or "text type" to pinpoint the exact item you should be looking for. Again, circle any subtle negatives you find, such as *not* or *except*.

7. Spot the Hedges

Words such as *almost, most, some,* and *sometimes* are words that are used in *hedging* language, which is language that denotes a claim rather than an absolute fact. Look for this type of hedging in questions and answers. Likewise, answer choices that assert something is *never* or *always* might need another look. Unless you know for sure, saying *always* or *never* may indicate an incorrect answer choice simply because of the absolutism of the term.

8. Don't Overanalyze

It's normal to be nervous while taking the test. However, be aware that with nervousness may come over-analysis. It's important not to read too much into questions and to avoid thought tangents about what the author could possibly mean behind what is actually said. If you find yourself overanalyzing or overthinking, shut your eyes, take a deep breath, and count to ten. Read the question again with a clear head so that you can answer with clarity rather than a muddled brain.

9. Don't Panic

Sometimes, you won't know the answer. In that moment, there will be nothing you can do to find the answer. To deal with this, compartmentalize each question; you may not know #27, but #28 is a new question that you are fully capable of answering. Leave the question anxiety with that question and move on to the next question. Make sure you close your eyes and take a deep breath before beginning again. This will calm your nerves. Say an affirmation, such as "I am fully competent to answer this question" or "I am calm and knowledgeable." An excellent way to understand a confusing question is to rephrase it yourself. What is the question asking you? Rewrite the question down on scratch paper and refer to the answers accordingly.

10. Retrace Your Steps

Mark any questions you know with absolute certainty right away. If you narrow the answers down to two choices, it might be worth the risk to choose one and move on. However, if you find yourself not knowing the answer at all, or having only narrowed the answers down by one (with several options left), move on. Leave the difficult questions for the end, if your exam allows you to go back and retrace your steps. One more tip; it's often prudent to go with your gut. So, don't change any answers that you've already marked as correct!

FREE Videos/DVD OFFER

We have created a set of Exam Tips Videos that **cover proven strategies that will teach you how to prepare for your exam and feel confident on test day.**

We want to give you access to these free videos as a token of our appreciation. All we want to know is what you thought of our product.

Here are the steps:

1. Email **freevideos@exampedia.org**

2. Put **"Exam Tips"** in the subject line

Add the following information in the body of the email:

3. **Book Title:** The title of this book.

4. **Rating on a Scale of 1–5:** With 5 being the best, tell us what you would rate our book.

5. **Feedback:** Give us some details about what you liked or didn't like.

Thanks again!

Introduction

Function of the Test

The Preliminary SAT (PSAT) is an exam given by the College Board intended to measure students' knowledge of learning as well as prepare them to take the SAT. The PSAT is taken in the 10th and 11th grades. The PSAT exam is given in schools across the United States.

Test Administration

The PSAT is offered during the fall in the month of October. Individual schools choose which date they give the test. Usually scores are improved after taking the PSAT a second or third time, which is why it is taken throughout high school. Students can check and see if their score matches their college's requirement. For accommodations offered for students with disabilities, check with your student's school.

Test Format

Regarding the testing environment, students should have a list from their school noting what they should bring on test day.

When testing begins, the coordinator will read all instructions and tell you when to start and stop on each section. Students may not go back to work on sections, and they may not start early on a section.

The PSAT includes the following sections:

- Reading
- Writing and Language
- Math

The Reading Test includes passages and multiple-choice questions. The Writing and Language Test has passages that provide grammar and stylistic mistakes students must correct using multiple-choice questions. The Math section contains multiple-choice and grid-in questions.

Scoring

Students can find their scores through their online score report. PSAT scores are sent to schools, districts, and states. PSAT scores are not sent to colleges. The score range on the PSAT is from 320 to 1520 points.

Recent/Future Developments

The Essay portion that is optional for the SAT is not provided on the PSAT.

Reading

Reading will play an inevitable role in both your college career and actual career. Regardless of the discipline you study or the industry you enter after college, you will be faced with a diversity of reading tasks. The PSAT Reading Test is designed to test a student's ability to interpret such a diversity of texts. Skills such as close reading and the ability to recognize both implied and stated meanings are drawn upon heavily throughout the test. You will also be required to locate evidence and other support for any claims an author may make, such as context clues. In scientific texts, the interpretation of data and a consideration of hypotheses and their implications may be necessary.

The test will further assess your skills in the following areas:

- Information and Ideas: What is the author's message? What details are presented in support of this message? Can you identify the theme of the passage you've been asked to read?

- Rhetoric: How does the author present his or her message? What are the author's opinions? How are the ideas structured?

- Synthesis: What connections can you make between two passages? Or, how do two passages differ in their ideas?

The three major subject areas covered in the PSAT Reading Test are as follows:

- U.S. and World Literature (e.g., Classic or Modern)

- History and Social Sciences (e.g., Sociology, Psychology, Economics, Historical Speeches)

- Science (e.g., Biology, Chemistry, Physics, Earth Sciences)

The test is presented in the following format:

Each subject area listed above is represented by a single passage or a pair of related passages. The passages range from 500 to 750 words in length. Each passage or pair of related passages is followed by corresponding questions, which will ask you to identify relationships between sets of passages or to think more deeply about a single passage. Some passages are paired with images or graphics, which you will need to view, read, interpret, and relate to the connected passage(s). In some cases, an image or graph may be critical to the meaning or purpose of the corresponding passage(s).

The complexity of the passages varies from familiar and straightforward to having higher-level vocabulary and layers of intended meaning. Just as genres vary, so, too, do the purposes of each passage. For example, some tell a story, while others are informational or present a process. Some are fiction, and others are nonfiction. It is up to the reader to identify the purpose of each reading, determine the rhetoric used by each author (e.g., language, tone, intended audience, potential bias), and then select the best possible answer choice for each question. One passage is literary based; one passage, or one pair of passages, is history based; one passage is social science related; and two are science based. The test includes approximately forty-seven multiple-choice questions, each with four answer options.

Command of Evidence

Finding Evidence in a Passage

As briefly mentioned earlier, some questions may ask you to locate **evidence** in a passage or passages that support an author's ideas. In other words, what does the author use to support his or her claims? Facts? Personal anecdotes? Are there quotations from other sources, such as an expert testimony?

Let's start from the beginning. What exactly is evidence? The list above (facts, anecdotes, quotations) contains types of evidence an author may use to support their claims. When an author makes a claim, they are making an assertion about something, perhaps to convince readers of a particular point of view or to propose a resolution to a problem. Whatever the agenda, the author will use evidence to back up their assertion. In other words, what does the author say, following the claim, to prove to readers that the argument is valid? To understand these ideas better, we'll first discuss how to locate a source's claims. Then, we'll run through the types of evidence an author may use to support their claims as well as how to identify each type in a reading.

A **claim** in a piece of writing is the text's argument. Everything that follows the claim should support the already established main point. Generally, a claim appears near the beginning of the piece of writing. Look in the first or second paragraph to locate the author's argument. An author typically makes three types of claims.

Claim of Fact

A **claim of fact** argues that something is true, has been true, or will be true in the future. For example, an author may use empirically collected data, such as scientific studies or historical documents, to back up such a claim. Here are some examples of claims of fact:

- Standardized testing in schools does not accurately measure a student's rate of success or understanding.

- Eyewitness testimony is often flawed and leads to more false identifications than other forms of police evidence.

- Torture, as has been used by the U.S. government against foreign adversaries, has not effectively produced valuable evidence in domestic and foreign terrorism cases.

Claim of Policy

A **claim of policy** argues that something should be changed or added in order to solve a problem or to improve something. Political candidates, for example, often use claims of policy to win votes during election season. Here are some claims of policy:

- The American health care system should be reformed in order to best meet the needs of working and middle-class families.

- To extinguish the illegal consumption of alcohol in off-campus student housing, the drinking age should be lowered to eighteen.

- All fifty U.S. states should place a ban on talking on cell phones while driving unless the call is made via a hands-free device.

Claim of Value

A **claim of value** argues that one thing or idea is better or worse than another. Generally, although these types of claims are based on the author's opinions rather than fact or policy, they are often used in a variety of contexts. Here are some examples of claims of value:

- The voices of Generation Z are far undervalued by generations who have come before them.
- Team sports teach invaluable lessons about respect, compassion, and how to work with others.
- The death penalty is a dehumanizing form of punishment in our current judicial system.

Once you have located the author's claim, you will want to look for evidence used to support the claim. The three most commonly used forms of evidence are as follows:

Facts

Facts are pieces of information that have been proven true, either by empirical evidence or observation. To support an argument, an author may draw on facts to prove to readers that the claims are founded in already-proven evidence. Take our claim of fact from earlier that eyewitness testimony has led to more false identifications than positive ones. To support this claim, an author may provide facts that show the number of cases in which a person of interest in a crime has been falsely identified and then later exonerated, albeit sometimes too late. In this case, the facts may appear in the form of statistics or numerical data collected over time. Facts will also be cited with the source.

Anecdotes

An **anecdote** is another word for *story*. An author may use a story to validate their experience or observation. For example, an author who argues that team sports teach valuable lessons might share a personal story to emphasize how that experience has positively impacted their life. A hypothetical story may be told to give readers an idea of how a situation might play out, such as a cause-and-effect scenario. It may also be used to draw attention to a topic by making the content more "realistic" to readers.

Quotations (Expert Testimony)

Quotations are the exact words of another person, enclosed with quotation marks. Typically, an author uses quotations to reflect the thoughts or ideas of someone else, usually derived from one of the author's sources. For example, an author may quote an expert in the subject about which they are writing. Doing so can increase the credibility of the argument if the quotation supports the claims being made in the reading. Let's say an author is arguing that torture is an ineffective form of eliciting information from suspects allegedly involved in a national security matter, such as an act of terrorism. The author may quote an expert in the field of domestic and international terrorism or national security. Quotations should be cited with the source.

Drawing Conclusions and Forming Inferences

Based on what you've read, you may also be asked to draw a conclusion or form an inference and then select the answer that best aligns with your own thoughts. **Drawing a conclusion** means determining the author's main point. What do they want you to take away from what you've read? Sometimes, drawing a conclusion requires you to read "between the lines," meaning you'll need to form an inference. **Inference** is another word for *conclusion,* but in this case, the ideas you may be drawing upon may not be implicitly stated in the text. Rather, they may be implied.

To draw a conclusion, consider using the following reading strategies:

- Read closely at the sentence level. Since you've already identified the passage's main point, think about what each sentence is saying about that main point. Take notes as you go.

- Next, read at the paragraph level. Look for threads. What does each paragraph say about the author's claim? What connections can you find between the individual paragraphs? Take notes on this analysis as well.

- Once you've read carefully at the sentence and paragraph levels and have identified the author's thoughts in each, look at your notes to find final threads. Has an idea been repeated? Does one idea, or multiple ideas, build on others? Finding these connections will help you draw a conclusion of your own.

Forming an inference from implied ideas requires a bit more detective work. When something is implied, it is not directly stated by the author. This means that the author assumes that, through the other directly-stated material in their work, readers should be able to figure out unstated connections or meanings. However, sometimes an author may unconsciously use language or a tone that implies a meaning they did not intend to present.

Context

Context is the situation in which language or behavior is set. When a person behaves or speaks a certain way, for example, they are likely being influenced by language or behavior that has occurred around them. For instance, when politicians give a speech, their word choice is dependent upon the political climate of the time. Thus, what they say is impacted by how their constituents (citizens) and other politicians are reacting to the most important issues facing them at that moment. This is context. In written work, context clues can be represented by word choice, tone, or any other hints a writer may give a reader to help them understand what is being said, such as expanded examples that may help to define a complex idea. Looking at word choice, tone, and surrounding hints may also aid in unearthing implied ideas. Here's an excerpt from a speech President Obama gave in 2016:

> If we turn against each other based on divisions of race or religion, if we fall for a bunch of "okey-doke," just because it sounds funny or the tweets are provocative, then we're not going to build on the progress we started. If we get cynical and just vote our fears, or if we don't vote at all, we won't build on the progress that we started.

In this passage, President Obama is arguing against passivity and cynicism in order to prevent a lack of continued progress. This idea is directly stated. He claims that inaction in terms of social media posts or the act of fear-driven voting will halt forward motion. However, he does not directly state what *should* be done. The answer? An intervention.

How do we know this? Look at the examples President Obama provides. He is very specific in terms of what behavior he believes we should avoid (e.g., turning against "each other based on divisions of race or religion," falling for "funny" or "provocative" tweets," voting by fear or not at all). Again, he argues that these behaviors will not allow us to "build on the progress we started." So, we should do the opposite—*not* turn against each other, *not* fall for funny or provocative tweets, and *not* allow fear to drive the act of voting. By *not* participating in these stifling behaviors, we can continue to press forward. Thus, we have an implied main idea.

Word Choice

Word choice is exactly what it sounds like: the words an author chooses to use in written work. In selecting words, authors may do so mindfully and purposefully, or they may do it unconsciously, allowing their unstated beliefs or feelings to appear on the page. Either way, the words an author chooses to present to readers can have an effect on how readers perceive what is written or how they act on what is written. Further, like context clues, word choice can help readers identify the main ideas in a written text, both directly stated and implied. Let's take a look at a speech from President George W. Bush in the aftermath of 9/11.

> We have seen their kind before. They're the heirs of all the murderous ideologies of the 20th century. By sacrificing human life to serve their radical visions, by abandoning every value except the will to power, they follow in the path of fascism, Nazism, and totalitarianism. And they will follow that path all the way to where it ends in history's unmarked grave of discarded lies.

In this passage, President Bush is speaking about the radicals who attacked the United States on September 11, 2001. Here, Bush clearly states that the attackers sacrifice "human life to serve their radical visions," thus clearly placing the people responsible in the camp of radicalism. However, we also have more than one implied statement when he lists the ideologies of "fascism, Nazism, and totalitarianism," as well as when he notes their demise in an "unmarked grave of discarded lies." These carefully chosen words imply the radicalism of the ideologies listed by identifying them as analogous to the 9/11 attackers. Bush also implies, through the use of his grave metaphor, that the ideas of the 9/11 attackers will be snuffed out, as were those of radicals who came before them.

Informational Graphics

Some passages are paired with informational graphics or images. Pay close attention to these. They are there for a purpose. It will be up to you to determine what that purpose is. You may need to identify how they support or relate to the corresponding passages. Here are some of the most common uses of informational graphics and images.

Many **informational graphics**, such as tables, provide readers with a summary of data that has been collected by the author or by a source used by the author. This data is usually presented in its raw form, meaning it has to be analyzed by the author and/or reader for it to make sense. In other words, you, the reader, are expected to make connections between the raw data presented in the table or figure and the information surrounding it. The point is for the table to serve as a visual representation of data that cannot be summed up by brief text. An example might be data that represents trends over time or emerging patterns, both of which may be too complex to express without a graphic.

Images, sometimes referred to as *figures* when paired with text, can take several forms, such as photographs, charts, graphs, diagrams, or maps. Each one carries its own purpose. A photograph, for example, is often used as a rhetorical tool, meaning it may be used by the author to incite emotion or provoke action from readers. Charts, graphs, or diagrams, on the other hand, can provide illustrations of relationships or ratios. Maps are often employed to show spatial relations. Below are more detailed explanations of informational graphics and images.

Tables

Tables consist of three main parts: the title, the column titles, and the body. A good table has a clear, concise title—one that identifies, up front, what the table represents. It is not uncommon for the title of a table to be referred to later in the text. The same characteristics of clarity and conciseness should underlie the column titles. Column titles give readers an idea of the elements or units of analysis that will appear in the body of the table. The body of the table is, of course, where the data actually appears, guided by the table's title and column titles. Below is a simple example of a table that might appear alongside text.

School Lunch Patterns 2014–2015		
State	Number of Students	Percentage Eligible for Free or Reduced Lunch
California	6,047,295	62.4%
Arizona	764,309	39.7%
New Mexico	639,492	43.8%

Images

As mentioned, when paired with text, an **image** is commonly referred to as a *figure*. Like tables, you may see an author refer to an image provided earlier in the text by using the title or number of the figure being referenced (e.g., Figure 4.1). Here are some strategies for analyzing common types of images used alongside text.

Photographs

As with reading, it is often helpful to first scan a **photograph** before really dissecting it critically. Think about the dominant impression you get when you view the photograph. Taking notes can help you keep track of your thoughts. Next, consider how the image relates to the author's thesis statement or relevant evidence. Does it support what has been said so far, or is the image a distraction? Once you've gained an initial impression of the photograph, look at it a second time but more closely. Do you see any text as part of the photograph or perhaps a caption above or below it? If so, what impact do the words have on the photograph? Do they help to convey a meaning more powerful than the image would by itself? Finally, consider the intention of the photographer. Look at the lighting, color, and juxtaposition of people or objects in the photograph. These are often purposefully used by the photographer to produce certain reactions in viewers.

Protests for Vietnam War, State Archives of Florida, Florida Memory, https://floridamemory.com/items/show/25387

Charts, Graphs, and Diagrams

Charts, graphs, and diagrams, as noted above, provide readers with visual representations of relationships or trends. Like tables, they also have main titles and titles of the units that are represented. Typically, they illustrate quantitative (e.g., numerical) values—ones that indicate how elements connect or impact one another. To accurately interpret data represented by graphs, charts, and diagrams, use the following strategies.

As with a table, it is important to first read the title of the graph, chart, or diagram. Identify what is being represented and which elements of that subject are being presented through data. An additional element you'll find on a graph, chart, or diagram is a key. A **key** is a tool that tells readers what visual elements, such as lines or colors, represent. See the example of a line graph below.

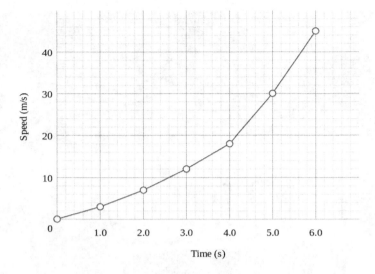

A chart, such as a pie chart, is another common form of graphic used alongside text. Charts often rely mostly on colors or shapes to help illustrate data. Like graphs, they have titles and unit titles as well as a key. In the example below, notice how the sections of various sizes represent the different elements identified in the key. Also notice how each section is labeled on the chart. The key will help you locate different pieces of data through strategies such as color coding or shading.

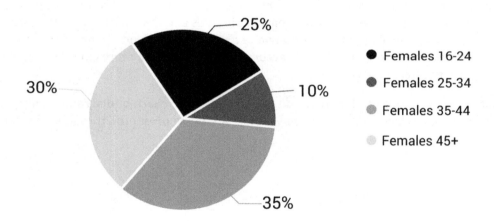

Connections in Graphics
Some graphics are used solely for the purpose of showing connections. For example, a **Venn diagram** is an illustration commonly used to show overlapping elements of two or more things as well as how they differ. Notice how each circle in the example below has, once again, a title and how the circles overlap.

The elements listed on the left and right are the differences, while the elements listed in the middle, where the overlapping occurs, are the similarities (e.g., connections).

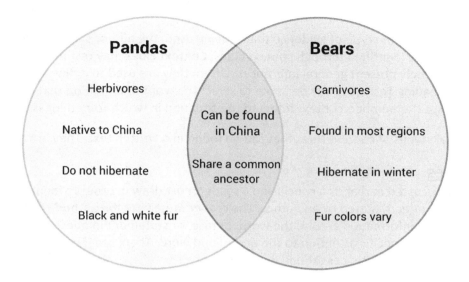

Maps

A **map** can take more than one form. For example, there are traditional maps, or ones that tell us where places can be found and how to get to those places. Then, there are mind maps, which guide readers through a person's thought process on a particular subject. Either way, the goal of a map is to locate something or, more specifically, to spatially represent where something is and how it moves from one place to the next. Here is an example of a map showing the world's largest religions and where they are generally practiced. Similar to tables, graphs, and charts, this mind map has a title and a key.

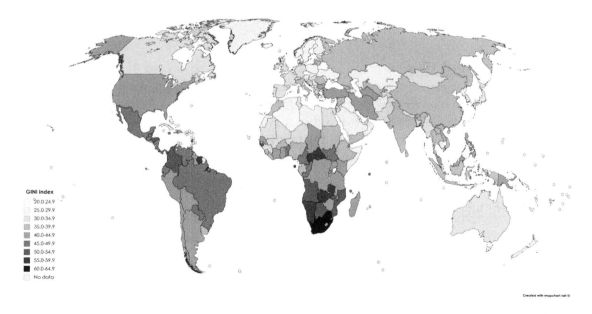

DennisWikipediaWiki; CC BY-SA 4.0, https://commons.wikimedia.org/w/index.php?curid=92529096

Words in Context

Context Clues

Sometimes, determining the type of evidence that is being used, the author's agenda, or whether the author is biased can be identified through context clues. **Context clues** may restate ideas, compare or contrast ideas, or simply present general inferences. Often, they are used to define words or phrases. As part of the PSAT Reading Test, you may be asked to select the word or definition that best fits the context of a passage. Remember, context refers to the situation in which something is set.

There are five main types of context clues that can be found in a written text. They are as follows:

Inference Clues

An **inference**, as discussed earlier, is a conclusion a reader must draw in order to identify an idea or meaning that is not directly stated by an author. The reader must *infer* the author's intentions or meaning by using the information around the word, phrase, or statement in question. Look at the example below, paying specific attention to the underlined word. Then, see if you can identify the word's meaning using the sentences around it.

> The new intern was starting his third week with the company, and he was not looking forward to it. His antipathy for the CEO was growing ever stronger. Each time she would enter a room, an involuntary snarl would spread across his face.

If you do not know what the word *antipathy* means, you may need to use context clues to identify its definition. We know that the intern was "not looking forward" to starting his third week with his new company, which means he was unhappy. In the third sentence, we are told that an "involuntary snarl" would spread across his face every time the CEO entered a room. Since a snarl indicates displeasure, we can assume the intern was displeased with the CEO. Thus, if he was unhappy with his internship and he expressed this unhappiness when he saw the CEO, a growing sense of "antipathy" must mean a growing sense of dislike.

Synonym/Restatement Clues

A **synonym** is a word that means the same thing as another word. Sometimes, an author may restate a word or phrase by using similar words with the same definitions. If you come across a word you do not know, look at the words around it to see if the author may have restated their idea in a different or simpler way. Keeping this strategy in mind, read the following example. Focus on the italicized word and try to determine its meaning by reviewing the words around it.

> I'd like to *capitalize* on your writing skills, taking advantage of your ability to market ideas to even the most stubborn clients.

Here, the word *capitalize* is part of a longer phrase, *capitalize on your writing skills*. We can assume, then, that *capitalize* is a verb, a word that seeks to invoke some type of action on a person's writing strength. The phrase, *capitalize on your writing skills*, is then followed by another phrase, *taking advantage of your ability*. From the sentence structure, we can conclude that the second phrase restates the first because to take advantage of something means to capitalize on it.

Antonym/Contrast Clues

An **antonym** is the opposite of a synonym. It is a word that means the opposite of another word. Like a synonym clue, an antonym clue can take the form of a word, a phrase, or an entire statement. To identify an antonym clue, look for words that indicate opposition or a shift in ideas. In the example below, focus specifically on the italicized term.

> Nikkita was expecting a *galvanizing* show when she attended the local theater production, but instead she was met with boredom and dismay.

In this sentence, one word in particular represents a change in thought patterns: *but*. The word *but* means that the writer is about to present an opposing idea—the idea that the theater performance was boring and dismal. This contradicts Nikkita's expectation that the show was going to be *galvanizing*. Thus, if the show was the opposite of her expectations (e.g., boring and dismal), we can assume she was hoping to see an exciting performance.

Definition Clues

A **definition context clue** is exactly what is sounds like—a word, phrase, or statement that defines a word, phrase, or statement that appears before it. Review the following example and try to identify the meaning of the italicized word.

> A common criticism of younger generations is that their members are often *materialistic*, focused primarily on their desire for things and status.

The italicized word, *materialistic*, is immediately followed by the phrase, *focused primarily on their desire for things and status*. The second phrase defines the word *materialistic* by providing more details. Sometimes, a definition clue begins with the word *which*, as in the example below.

> The new pathology assistant was directed to *gross* his first tissue sample, which means he was asked to inspect it without the aid of a microscope.

The word in question is *gross*, and it is clearly defined by the phrase, *which means he was asked to inspect it without the aid of a microscope*. Thus, if you see a connecting word such as *which*, read carefully, as it may lead you to the answer you are seeking.

Here are some additional strategies for identifying definitions, or meanings, using context clues.

Read the entire passage: Before you start guessing the answer to the questions that have been presented, be sure to read the corresponding passage, or passages, in their entirety. Since context refers to setting, whether physical, sociological, political, or other types, it is important for you to understand what is being said before you jump to conclusions.

Predict what you think the answer is: Once you've read through the passage or passages, see if you can identify the answer without looking at the options. This can be a valuable strategy because sometimes the answer choices can make you question your own thoughts.

Swap out answers: If you're still unsure about which option is correct, try each answer out to see which one makes the most sense. This is not a fail-proof strategy, but it can help. Be sure to still pay close attention to the context clues surrounding the word, phrase, or statement in question.

Read, read, read: The best way to improve your own vocabulary is through reading. Read all sorts of things, from fiction to nonfiction and poetry to newspaper articles. This will expose you to a variety of writing styles and vocabularies that you may have never experienced before.

Word Choice

Word choice can be critical to a reader's understanding of a passage. The words an author uses can be a window into their unstated thoughts, such as biases, and can indicate the author's tone and intended or unintended meanings. Read carefully.

Moreover, words can be used to paint a picture—one that can appeal to a reader's senses or emotions. Carefully chosen words can tell a story in which readers are met with smells, sights, sounds, or feelings that draw them into the setting of the text, right alongside the characters or narrator. In addition, sometimes an author specifically chooses language that will incite a particular feeling in a reader, such as anger, sadness, or joy. This is often done in persuasive writings or speeches in which a writer or speaker seeks to gain the support of the audience.

Below are some ways in which writers use language in their texts, both intentionally and unintentionally.

Denotation and Connotation

The **denotation** of a word is the exact dictionary definition. **Connotation**, on the other hand, is the meaning associated with a word—one that may lead readers to think of other words with other meanings. In the following example of a word's denotation and common connotations, focus on the italicized word.

> In the Appalachian region of the United States, a significant portion of the population is living in *poverty*.

If we look up the dictionary definition of the word *poverty*, we'll find that it means lacking resources, such as money or goods. We may find a definition similar to "the condition of being poor." According to the U.S. government, this definition may be even more specific, associating poverty with an actual dollar amount in terms of annual or monthly income.

However, if you write down your own ideas about poverty, such as words or descriptions that come to mind without consulting a dictionary, you may find yourself identifying meanings that are extraneous to the word's actual definition. For example, common connotations associated with the word *poverty* may include *dirty, disheveled, hopeless, unemployed,* or even *homeless*. For those living in poverty, however, these descriptions may or may not actually be accurate. These are only our connotations of what the word *poverty* means.

When reading a text, therefore, it is important to identify the meanings and/or descriptions being implied or directly stated by the author. Think: Does the word being used actually mean what the author is saying it means, or is the word the author's, or society's, connotation of the word?

Jargon

Jargon is language specific to a particular field of study or industry. As a reader, you will typically find jargon in journal articles or papers associated with a field or industry's subject matter, such as an article in a medical journal or an industry's trade publication. Here is an example of a statement using jargon specific to the field of education.

> Today, many educators are utilizing the flipped classroom method of teaching in order to initiate critical-thinking skills and a growth mindset within their students.

If you are not an educator, you may find yourself wondering what a flipped classroom is or what it means to think critically. And what exactly is a growth mindset? The bottom line is this: Writers who employ jargon in their texts assume readers know the terms they are using because they assume their audience is in their field or knowledgeable about their field. Thus, as a reader who may not be familiar with the field being discussed in the text, you may have to use context clues to help you determine the jargon's meaning. This is especially important during a reading test, as you will not be able to look up the meanings of the terms being used.

Colloquialisms

Colloquialisms are words or phrases that are local to a specific region of a country. If you come across a colloquialism in a piece of writing, it is likely that the author or the characters are from a particular area of the country. Generally, the use of colloquialisms can be found in more informal or fictional writing, such as correspondence, personal journals, blogs, poems, and stories. For instance, some examples of colloquialisms that are often attributed to the Southern United States include: won't amount to a hill of beans (e.g., something is worthless), bless your heart (expressing thanks, sympathy, or even an insult), 'til the cows come home (something lasting a long time), over yonder (over there; far away).

Like jargon, if you are unfamiliar with a region's colloquialisms, it may help you to lean on context clues to identify an author's meaning.

Loaded Words, Phrases, and Statements

When an author uses **loaded language**, they are using words, phrases, or statements that may hold layers of meaning, bring up many questions, or incite a significant level of emotion. Often, this type of language is used to help readers understand how a narrator or character is feeling (e.g., appealing to a reader's senses or emotions) or to persuade readers to support an author's argument or cause. Sometimes loaded language is referred to as *emotionally charged language*. Think of an electrical charge. When there is an electrical charge, a large amount of energy is being distributed. The same can be said for words. Words can emit an enormous amount of meaning (e.g., energy), which can lead readers to perceive a message or act on a message in a way that has been influenced by the language being used. Thus, as a reader, it is important to read carefully because sometimes a writer may attempt to use loaded language to deceive an audience.

Below is an excerpt of a 1940 speech given by the then-leader of the Italian National Fascist Party, Benito Mussolini. Notice the words he chooses to encourage support for his belief system (Fascism) and subsequent entrance into World War II.

> Fighters of land, sea and air, Blackshirts of the revolution and of the legions, men and women of Italy, of the empire and of the Kingdom of Albania, listen! The hour destined by fate is sounding for us. The hour of irrevocable decision has come. A declaration of war already has been handed to the Ambassadors of Great Britain and France. We take the field against the plutocratic and reactionary democracies who always have blocked the march and frequently plotted against the existence of the Italian people.

Here, Mussolini lays out many carefully chosen words that seem to imply that other nations are plotting against Italy. His use of the words "revolution," "destined by fate," "irrevocable decision," "declaration of war," "and "plotted against the existence of the Italian people" are all examples of loaded language. His goal is to persuade listeners to believe the same ideas he believes and, subsequently, allow him to make, without opposition, the political decisions he desires to pursue.

Of course, most passages you read do not attempt to persuade their audience to adopt a fascist ideology, but nonetheless, authors of many belief systems, backgrounds, and educational levels use loaded language to some degree to gain support for a cause or help their readers feel a certain way about their topic. It is therefore important for you to take note of how an author's choice of words makes you feel or think about what you are reading and determine if their use of language is truthful or deceptive.

Analysis in History/Social Studies and in Science

Examining Hypotheses

As you've likely learned in your science classes, a **hypothesis** is an educated guess. In the scientific passages presented, you may be required to analyze an author's hypothesis. What exactly does it mean, however, to examine a hypothesis? To understand more clearly, let us back up for a moment. Generally speaking, the format of a passage that includes a hypothesis looks something like this: a problem statement or research question, a hypothesis, an explanation of a study that has been conducted, and a conclusion.

A **problem statement** or **research question** expresses the problem the author is attempting to solve or a question they are attempting to answer. The hypothesis can be thought of as a prediction about why or how something has occurred or will occur, or the author's best guess about the answer to the research question. To put it another way, it is an educated guess about how a study that has yet to be completed will conclude. The hypothesis is followed by an explanation of the study, usually including theories and/or methodologies used by the person completing the study as well as a discussion of the variables involved. At the end of the passage, the author provides a conclusion about what actually happened and whether or not their hypothesis has been validated.

Here is an example of a research question and a corresponding hypothesis.

> Question: Does the number of laboratory sessions attended by first-year biology students at XYZ University have an effect on final test scores?
>
> Hypothesis: The number of laboratory sessions attended by first-year biology students at XYZ University has a significant effect, both positive and negative, on final test scores.

Following this question and hypothesis, the author will likely discuss the way in which they went about validating, or not validating, the hypothesis. Pay close attention to the steps the author took in identifying which variables they chose to include in the study, the theories that underlie the study, and the methods used to validate the hypothesis. For example, if you were presented with the question and hypothesis above, you would want to understand why the author chose to study this topic, why they chose to study first-year students at that specific university, how students were chosen for the study, and how the data was collected to test the hypothesis. It is important for all of these steps to make sense and be justified, or the study may invalidate itself before it begins.

Here are some additional questions and strategies that may guide you in effectively reading a scientific study, which will in turn help you more accurately examine a hypothesis.

- First, why is the author studying the topic? Read the study's introduction and background (if included) carefully. Understanding the reasons the author has conducted a particular study will help you better understand their hypothesis and subsequent conclusion.

- Next, what exactly is the author trying to determine? Write the author's hypothesis in your own words. Being able to summarize another writer's ideas in your own words is a good test to determine your level of understanding.

- Then, look for the author's actions. What was done? How did the author go about answering the questions? What data was collected, and why? Being able to clearly identify strategy is key to understanding whether or not an author's conclusion logically follows the methods.

- Did the author adhere to a particular theory or set of theories? A theory is a guiding belief that may or may not play a role in the approach an author takes to the study.

- After examining the explanation section of the study, read through the results and discussion sections. What does the author conclude about the study and his or her hypothesis? Was the hypothesis validated? Why or why not?

- Remember to look carefully at any graphs, charts, or other images included with the passage. These often represent a visual display of the study's data.

- After you've read through the entire study, go back and reread the introduction. Was there anything you missed? Was there perhaps something you misunderstood, which may have led to a subsequent misunderstanding of the study's conclusion? If not, does the conclusion follow the introduction in a logical manner?

Reading and understanding a scientific passage can be challenging, but taking it step-by-step and reading each section carefully can make a big difference in your understanding of the author's initial intentions and conclusions. Think of it as a great way to practice your critical-reasoning skills.

Interpreting Data

Scientific passages may also include data that will need to be interpreted so that you may draw reasonable conclusions or inferences. Typically, the data is presented in graphs or charts, and you will be asked to determine how the data relates to the corresponding passage(s). It is important for you to read the passage, or passages, very carefully first and then do the same for the chart or graph. You will be tested on your ability to understand the data represented by the chart or graph.

Remember that a chart or graph is labeled and may include a key. Before you attempt to answer any questions about the data, look closely at each element. What is the title? What are the headers? Which elements of the topic are being interpreted by the chart or graph? If there is a key, what does the key represent, and how (e.g., colors, lines, bars)? Look at the example below.

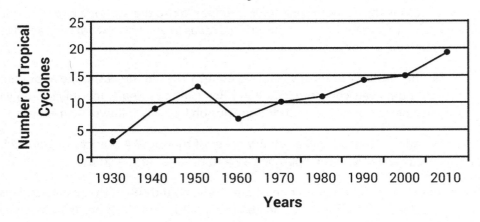

Notice that there are three labels: the graph's title (Number of Tropical Cyclones Every 10 Years), the element being analyzed (Number of Tropical Cyclones), and the element by which the number of cyclones is analyzed (Years). The cyclones are then plotted on the graph using a line and dots. The line represents the rise in cyclones over the course of several years, while each dot represents the year interval at which a cyclone occurred.

Here is a sample question that tests a reader's ability to read the graph.

Approximately how many tropical cyclones occurred in the year 1940?
    a. 5
    b. 9
    c. 11
    d. 13

To answer this question, you'll first want to read it carefully. Determine exactly what the question is asking. You may want to go back and reread the passages once more to ensure you correctly understand the data. Sometimes, the passage includes a verbal explanation of the data, which may provide readers with more clarity. For example, if the number of cyclones that occurred in 1940 was not exactly identified by the graph, as it is here, you may need to approximate the correct answer by looking carefully at how the storms are plotted on the graph and comparing this information to information discussed in the passage. Using the text and graph together is important to drawing more accurate conclusions or inferences.

Other questions will test your ability to determine the validity of a statement or the answer choices presented. Here's an example.

Based on the graph above, which of the following statements is true about the difference in storm activity between the years 1930 and 2010?
   a. The storm activity gradually increased every 10 years between 1930 and 2010.
   b. The storm activity remained at a steady level every 10 years between 1930 and 2010.
   c. The storm activity between 1930 and 2010 decreased over the course of 10-year increments.
   d. The storm activity increased sharply every 10 years between 1930 and 2010.

If we look back at the graph, we can determine that between 1930 and 2010, based on the slight uptick in the numerical data, the storm activity gradually increased. However, if the data was perhaps less clear, or harder to distinguish, you could try to find the correct answer by first eliminating incorrect ones. Look for answers that are *not* validated by the data in the graph or the passage, and eliminate them from your choices. Continue this strategy until you are able to identify the most likely answer.

For example, between the years 1930 and 2010, there was only one instance of decreased storm activity in the year 1960. Thus, we can eliminate the answer option, "The storm activity between 1930 and 2010 decreased over the course of 10-year increments." We can also eliminate the answer option, "The storm activity increased sharply every 10 years between 1930 and 2010." Although the number of storms increased every 10 years, the numbers did not rise sharply. We can also eliminate "The storm activity remained at a steady level every 10 years between 1930 and 2010" because if it gradually increased, it couldn't have remained steady.

Let's look at another example. The bar graph below is a figure depicting the amount of physical activity recommended each day for four different age groups. Review the graph carefully, including headers, labels, and the key, and then consider the question that follows.

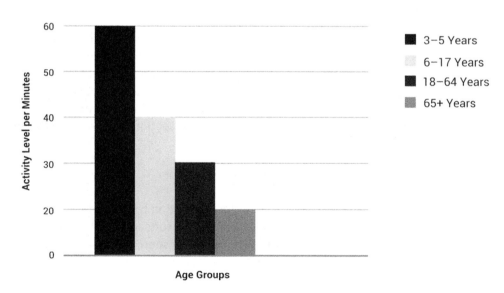

Based on the bar graph, which age group should be getting the most daily physical activity?
   a. 3–5 years
   b. 6–17 years
   c. 18–64 years
   d. 65+ years

If you can answer this question using the data presented on the graph, it is a good indication that you are able to accurately read a bar graph. We can tell by looking at the height of the bars that the purple bar is the tallest, reaching 60 minutes per day. According to the key, the tallest bar represents 3–5 years. Thus, the age group 3–5 years, according to this hypothetical graph, requires the most activity each day.

Here's another question based on the same bar graph.

Which of the following statements is true about two of the represented age groups?
   a. The age groups 3–5 years and 6–17 years do not require much daily physical activity.
   b. The age groups 6–17 years and 18–64 years require the most daily physical activity.
   c. The age groups 18–64 years and 65+ years require equal amounts of daily physical activity.
   d. There are no overlapping similarities between any two of the four age groups represented on the graph.

The answer here is, "The age groups 18–64 years and 65+ years require equal amounts of daily physical activity." We know this because the two colored bars that represent these age groups appear at equal intervals on the graph.

By using simple checkpoints, such as headers, labels, and keys on graphs and charts and pairing this information with hypotheses, explanations, and conclusions discussed in the corresponding passage, you will find it much easier to interpret data and draw necessary conclusions.

Considering Implications

Implications can create layers of their own meanings. You may be asked to identify or analyze implications in scientific passages based on your understanding of the information. Like inferences and conclusions, implications can be stated or unstated interpretations of information; they are connections that are drawn out of the information that is provided, even if the evidence is not complete. At times, implications can also formulate a discussion of consequences that have been made. In scientific research, implications are most commonly found in the discussion section (e.g., the conclusion) of a study.

Reading for Implications in a Scientific Study

When reading a scientific passage, you will want to pay close attention to the author's conclusions. In this section, they may discuss (or imply) implications of a study's results. Implications may include a discussion on how a study's findings may be important to further research or how it should affect future actions or policies. As noted earlier, implications may also identify, either directly or indirectly, potential consequences of a study's results. Here are some questions to ask when reading the discussion section (e.g., the implications) of a scientific study.

- What conclusion is the author drawing?

- Does the conclusion answer explicit questions? Do the answers to these questions align with your own interpretation of the study?

- How is the author interpreting the results of the study?

- Do you agree with the author's discussion of the results, or do you interpret them differently? You may want to read the passage again, if time allows, to reevaluate your understanding.

- Do you find any faults in the author's discussion section?

- Does the author include any graphs or charts alongside their study? If so, in what way do these graphs or charts help you understand the data produced by the study?

- What does the author propose to do next? Does he or she propose action, further research, or both?

Practice Quiz

Questions 1–5 are based on the following passage:

> Among many other factors, the risk of adult obesity is greater among adults who were obese as children, with racial and ethnic disparities existing by the age of two. If nothing else is done in the United States beyond what is being done now, simulated growth trajectories that model today's children show that over half (59% of today's toddlers and 57% of children aged two to nineteen) will be obese by age thirty-five. Early feeding patterns, including how babies are fed and how caregivers use food in response to an infant's mood, affect acute growth, future eating patterns, and the risk of obesity. Similarly, family and caregiver modeling of healthy behaviors, food offerings, and active playtime, as well as characteristics of neighborhoods such as walkability and traffic volume, may affect children's nutrition and physical activity habits.
>
> As sectors come together to reduce the obesity epidemic, we are aware how challenging success will be due to factors such as 1) the contributing risk factors of genetic and biological attributes; 2) individual behaviors (parenting styles, dietary patterns, physical activity levels, medication use, sleep, stress management); and 3) community and societal factors that influence individual, family, and collective access to healthy, affordable foods and beverages; access to safe and convenient places for physical activity; and exposure to the marketing of unhealthy products.
>
> By using self-reported data of height and weight from the Behavioral Risk Factor Surveillance System, CDC's Division of Nutrition, Physical Activity, and Obesity (DNPAO) has published state-specific obesity maps since 1999. Obesity is defined as a body mass index (BMI: a person's weight in kilograms divided by the square of height in meters) of 30.0 or higher. These maps have shown the growing epidemic that has affected our nation from coast to coast. Although the data collection methods changed in 2011, which somewhat limits our ability to assess trends, the 2017 data continue to show that obesity prevalence among adults remains high across the country. The state-specific prevalence ranges from a low of 22.6% in Colorado to a high of 38.1% in West Virginia.
>
> Petersen R, Pan L, Blanck HM. Racial and Ethnic Disparities in Adult Obesity in the United States: CDC's Tracking to Inform State and Local Action. *Prev Chronic Dis*. 2019;16:E46. Published 2019 Apr 11. DOI:10.5888/pcd16.180579

1. If we don't change our tactics to address obesity, which of the following is true:
    a. Obesity levels in adults will level out.
    b. As long as they weren't obese as children, adults will be fine.
    c. Over half of American children will be obese by age thirty-five.
    d. Children of obese parents will likely be obese as well.

2. The author writes that there are many challenges to tackling obesity, included among them are individual, community, and societal behaviors. What other factors contribute to the challenge?
    a. Fast food
    b. Biological and genetic attributes
    c. Science of the food industry
    d. Paying to play sports

3. Increasing this feature in a neighborhood might have a positive impact on a child's nutrition and physical activity habits:
   a. Single-family homes
   b. Traffic
   c. Convenience stores
   d. Walkability

4. How is obesity defined?
   a. Weight
   b. Size
   c. Appearance
   d. BMI

5. "Among many other factors, the risk of adult obesity is greater among adults who had obesity as children, with racial and ethnic disparities existing by the age of two." Based on context clues from the passage, what does the word *disparities* mean?
   a. Differences
   b. Experiences
   c. Beliefs
   d. Similarities

Answer Explanations

1. C: According to the text, "If nothing else is done in the United States beyond what is being done now, simulated growth trajectories that model today's children show that over half (59% of today's toddlers and 57% of children aged two to nineteen) will have obesity at age thirty-five." Choices *A*, *B*, and *D* are not true.

2. B: There are three primary challenges identified. The first challenge is genetic and biological components. While Choice *A*, fast food, might contribute, it is not one of the factors mentioned in this study. Though they might be part of a larger discussion on the issue, neither Choice *C* nor Choice *D* are discussed in this specific article either.

3. D: Increasing walkability in a neighborhood might have a positive impact on a child's nutrition and physical activity habits. In fact, the passage notes that traffic, Choice *B*, is a problem. Choice *A* and Choice *C* are not mentioned.

4. D: According to the passage, "Obesity is defined as a body mass index (a person's weight in kilograms divided by the square of height in meters) of 30.0 or higher." Choices *A*, *B*, and *C* do not correctly define obesity.

5. A: The first half of the sentence suggests a reason why we might see different obesity rates in specific children. The second half of the sentence, by way of comparison, suggests that racial and ethnic demographics have an impact as well. Therefore, *disparities* means *differences* so that researchers can account for why some children have a greater chance of becoming obese (parental, racial, ethnic differences). Choices *B*, *C*, and *D* are not appropriate substitutions for the word *disparities*.

Writing and Language

Writing will be critical during your time in college. It will also be critical to most career paths you may choose following college. From taking notes to writing essays in your college classes, to sending emails and creating reports while in your job of choice, the need to write clearly and effectively can be found in a multitude of disciplines and contexts. Effective writing cannot simply occur, however, without the writer taking the time to proofread and edit their own work. Proofreading entails ensuring that grammar, spelling, and capitalization are correct and communicate the writer's true intentions. Editing, on the other hand, requires a deeper look into the writer's rhetorical strategies, organization, argument, and support. Throughout your time in college, you will likely learn and practice both proofreading and editing skills to improve your own writing. The Writing and Language test will be the first step in assessing your current ability to analyze and revise written work. To do so, you will be presented with a variety of passages written specifically for the test. This is done intentionally so that strategic errors may be added.

Passages

The passages on the Writing and Language test vary in complexity, purpose, and discipline. Some use vocabulary and sentence structure that are simple and straightforward, and others use more complicated vocabulary and sentence structure. Some seek to convey a story, using narration as a strategy, whereas others may argue a point or present content for informational purposes only. The disciplines covered are Humanities (art, music, film, literature), Science (biology, chemistry, geology, physics), Social Studies/History (anthropology, social sciences, psychology, economics), and Career Studies (discussions on job trends and issues). Each passage is approximately 400 to 450 words in length.

Question Types

Each passage corresponds with a set of questions. The questions fall into one of two categories: expression of ideas or standard English conventions. Questions that cover the expression of ideas will ask readers to review an author's use of rhetoric. This may include analyzing an author's organization, the effectiveness of an argument, the use of (or lack of) supporting ideas, transitions, language choice, and introductions and conclusions. Questions that cover standard English conventions will ask readers to look at an author's use of grammar and mechanics. More specifically, these questions will test your ability to recognize errors in grammar and mechanics and select the best revision.

Format

The format of the test will appear as follows: A passage is presented on the left-hand side. The questions that correspond with the passage appear to the right. Some questions correspond to a specific place in a passage; if this occurs, sentences in the passage are numbered. This same number is presented next to the question about that sentence.

Directions may be specific to a concept or numbered point in a passage. If so, read carefully and follow the guidelines presented to you for completing that question. If directions do not appear, assume that you are being asked to select the most rhetorically effective addition or revision to a passage or the grammatically correct answer.

Command of Evidence

Using Evidence to Support Claims

Once you've identified the type of claim presented by an author, and any directly stated or implied ideas, it's important to notice how the evidence used in the text supports ideas in the passage. Keep in mind the types of evidence discussed earlier: anecdotes, facts, and quotations.

Remember, the point of using evidence is to support an assertion the author is making. This assertion can be an argument that something is good or bad, a proposal for a solution to a problem, or even a lesson learned from experience. Once you've identified the author's evidence, ask yourself: Does this evidence move the author's assertion forward? Or does it stall the argument in one place? Evidence should always help readers learn something new or understand something in a different way. If the evidence provided does not do this, it may not be important to the author's claim. For example, if an author is writing an essay about voter oppression in the 21st-century United States, it is important to provide historical examples of how this issue propagated itself into American society in the first place. This would be important to help the audience understand the current claims the author is making by allowing them to see the argument's foundation.

Evidence in Relation to Thesis

Just because evidence or text relates to the author's claims does not necessarily mean it is important information. For example, evidence may relate to the assertion being made, but it may not help move an argument forward. Or, a piece of evidence may be an important idea, but it may not directly connect to the main point of the argument. As a reader, it is important for you to clearly understand the author's main argument before analyzing the evidence supporting it. Truly understanding the claim will help you identify any irrelevant evidence in the text. Evidence must also come from a credible source, meaning a source that is trustworthy and accurate.

Here's an example of irrelevant evidence. Let's say you come across a passage about homelessness in major U.S. cities. The author argues that there should be a partnership between government and private entities in order to effectively make progress on this issue. However, the author then provides evidence that supports an increase in funding for local public schools so that more resources may be extended to all children. While this evidence is important and may help improve the lives of homeless children, the funding of public education is an entirely separate argument.

Implied Meanings

As discussed earlier, anything an author includes in their text may have directly stated or implied meanings. This idea is relevant to any evidence the author provides. Why that anecdote, for example? Why those statistics? Is there an implied meaning or reason the author may have chosen to use that evidence as support? For instance, one common tactic used by authors is to present readers with a startling fact or statistic in their introductory paragraph. The use of such a tactic may imply that the author believes one way or another about the topic of the fact or statistic. Let's look at the following example of an introductory paragraph that uses a statistic to attract readers' attention.

> Individuals who have dogs as pets live 28 percent longer than those who do not have dogs. Dog owners tend to get more exercise and enjoy companionship not typically found with other animals. Owning a dog may also provide people with the drive to care for someone other than themselves.

Without directly stating it, the author is implying that owning a dog is beneficial to humans for multiple reasons. Using the statistic of "28 percent" in relation to life expectancy also implies that dog ownership has a significant impact on a person's health. Thus, this statistical evidence is used to present an argument in a way that does not need to be directly stated in order to gain a reader's attention.

Author's Connections

An author may also use evidence as a way to draw connections between ideas or between ideas and real-life examples. Just as we did when looking for main ideas, we can also locate threads between sentences and paragraphs. Anecdotes, for example, may be one way an author directly connects a claim to an event that occurred in real life. Such connections may help a reader understand the author's claims in a clearer or more relevant way.

A common argument, for example, is that drug trials for humans should not be conducted using animals. If an author made such a claim, they may then use anecdotal evidence to connect the assertion to real-life issues. In this case, the author may present a story of inhumane treatments given to animals in laboratories during the course of a drug trial as well as the aftereffects of the treatments. Such evidence may help the author make connections between their assertion and real things that have happened to real animals in real laboratories.

How Evidence is Used

Finally, evidence is most often used to expand upon an author's claims. This expansion may take the shape of detailed examples, definitions, support by experts in a particular field, or as observed or empirically-collected data. The way an author chooses to use evidence can be the reason a reader understands, or does not understand, an author's assertions. It may also be the reason a reader agrees or disagrees with the argument.

If, for example, an author claims that working from home actually increases the productivity levels of employees, they may use the expert testimony of a workplace productivity specialist to support that claim. Providing opinions from experts in a relevant field may make an author's argument more credible for readers, thus increasing the likelihood that they will take the argument seriously.

Words in Context

When taking the PSAT Reading test, you are asked to analyze how an author uses language. On the Writing and Language test, however, this skill set is taken a step further. You will be asked to first consider the author's word choices, and then you may be asked to make choices about how best to use language when revising a written text. For example, you may be asked to select a word or phrase that best expresses the author's meaning or to choose the best revision for a wordy sentence. Below are discussions of some of the areas you may be asked to consider.

Effective Language Use

Effective language use entails not just word choice but also precision, conciseness, style, tone, mood, consistency, and syntax. In other words, it is not just the words a writer chooses to use that is important but also how a writer uses those words. These choices will be impacted by the writer's audience, the purpose behind the writing, and the level of formality expected for the audience and purpose.

Precision

Having **precision** means carefully selecting words that mean exactly what you are trying to say, or as close to your intended meaning as possible. Remember, words can have both denotative and connotative meanings. A writer should know a word's dictionary (denotative) meaning as well as its possible associations (connotations) and use the word in a way that ensures the intended interpretation. Let's say, for example, that you are trying to represent the way a person is smiling, but you are not sure which word to use. However, you know the person is not being sincere in their smile. You try two different words within the same sentence:

> The smirk on her face told me that her comments were anything but genuine.
>
> The sneer on her face told me that her comments were anything but genuine.

Both words, *smirk* and *sneer*, mean a smile that is not sincere. However, one word, *smirk*, means the smile of a smug or arrogant person, whereas *sneer* means a smile marked by contempt or disdain. The first emits confidence, however misguided, and the second reflects scorn. Knowing which word accurately expresses the true feeling of the person smiling will guide the entire meaning of the sentence. If not chosen correctly, this guidance will be a misguidance.

Knowing the expected audience for a piece of writing can also help an author choose precise words; if an audience, for example, has a **bias**, or a particular belief about the subject being discussed, they might also understand the terminology related to that subject. Having an awareness of this in advance can help a writer either avoid furthering the audience's bias or help the writer persuade readers by using language they know will appeal to them. Two hot topic words right now, for example, are *immigrant* and *alien*. Both words mean a person who is not originally from the country in which they live. Neither word is defined by negative language or derogative connotations, but based on the audience to which the words are written or spoken, a clear bias may be attributed.

For instance, if an audience is in favor of supporting migrant communities, they will most likely view the word *immigrant* as having a more positive connotation. However, an audience with a more challenged view of the current immigration system may be more comfortable with the word *alien*, which carries a more negative association. Although both audiences feel their word is more on point because of the societal weights they carry, denotatively, they both mean the same thing. Nonetheless, an author may choose one word over the other because they know the intended audience will react to it. Alternatively, an author may choose to avoid words that could potentially incite an audience in order to bring some objectivity to the discussion.

Finally, sometimes a word needs additional information to help readers understand the author's usage. It may not, for example, evoke an immediate denotative or connotative interpretation from readers.

Instead, a word may be too vague or subjective to project a clear meaning on its own. Here is an example. Keep your eye on the italicized word.

> Dr. Jacob Stephenson was an *interesting* individual, one who stirred up quite a bit of curiosity and confusion in me.

If we look up the word *interesting* in the dictionary, we might find definitions such as *capable of engaging a person's attention*. But what exactly does that mean? Engaging one individual's attention may not do the same for a different individual. Thus, the writer must better define the word *interesting*. This can be accomplished by giving detailed examples of the behavior or personality traits the writer deems interesting. The more specific and on-target a writer can be with their choice of words, the more on-target they will become with the audience's interpretation of the written work.

Conciseness

Whereas being precise means choosing the most accurate word for the intended meaning, **concision** means saying what needs to be said with the fewest words. To do this, a writer must select words that are the most direct; the more direct the words, the fewer words that are needed. Sometimes, during revision, an author may need to make one or more sentences more concise because they are too wordy. Take a look at the following examples.

> Before we decide which vacation to take this summer, we should consider, review, compare, and contemplate each location's tourist attractions, historic sites, and food and music scenes.

That is quite a lot of words to say something so simple. In the context of the sentence, the words *consider, review, compare,* and *contemplate* mean roughly the same thing, so why not select just one? Furthermore, *tourist attractions, historic sites,* and *food and music scenes* are also similar in their category of meaning. Only one is necessary. Here is a revised version of the same sentence.

> Before we decide which vacation to take this summer, we should compare each location's tourist attractions.

Notice how much shorter the revised sentence is compared to the original. By removing repetitive language, we were able to express the same thought in a more straightforward way.

Another way to make writing more concise is to remove words that serve no purpose. Again, these words can often be replaced by fewer words that get straight to the point. See the example below.

> In spite of the fact that the patient requested more medication, the doctor did not prescribe him additional painkillers.

Let's focus on the following phrase: *in spite of the fact that*. Most of those words could actually be eliminated. To achieve the same meaning, we could replace them with singular words, such as *despite* or *although*. Here is that same sentence again, this time more concisely written.

> Despite the patient requesting more medication, the doctor did not prescribe him additional painkillers.

We have taken six words, *in spite of the fact that*, and morphed them into one, *despite*. In doing so, the reader engages with fewer words to achieve the same goal: understanding the author's point.

Style, Tone, Mood, and Consistency

Style

In writing, **style** is subjective, just like preferences in fashion, music, or even food. Some writing styles may appeal to one group of readers, whereas those same styles may come across as awkward to another group of readers. For example, a sentence may be grammatically correct in every way, but to some readers, it may also be unnecessarily lengthy. Or, a writer may be prone to splitting infinitives, which is placing an adverb or adjective between the infinitive *to* and its verb—for example, *to suddenly awaken*, *to boldly go*. The latter is a habit that was once frowned upon but is now more widely dismissed as an author's style. However, some people still dislike this habit and consider it a grammatical error. The most important thing to consider when revising for style is to determine whether the style in question interferes with the author's meaning. If it does, for the reader's sake, it may be beneficial to revise it. Below is a brief discussion of three areas in which style sometimes causes a hindrance to an author's message.

Clichés

A **cliché** is a well-known expression that is overused. Because it is overused, its meaning and impact become less original. As a result, readers may skim over clichés in a piece of writing and might even become annoyed by them. To avoid this, it is far better for a writer to say, in their own words, what is really meant by the given statement. Or, a writer may choose to remove the cliché altogether. Here are some examples of clichés and possible revisions. In each sentence, the cliché is underlined.

> Cliché: As the old saying goes, time heals all wounds.
>
> Revision: It is a commonly believed that, over time, a person will heal from negative experiences.

Notice that in the example above, the cliché has been rewritten in the author's own words. As such, the statement becomes clearer and more interesting to read.

Here is another example.

> Cliché: Scientists became very excited about these new findings, just like a kid in a candy store.
>
> Revision: Scientists became very excited about these new findings.

Here, instead of rewriting the cliché, the author has completely deleted it. Because being *excited* means the same thing as the cliché *like a kid in a candy store*, it is unnecessary to include both.

Passive Voice

Passive voice occurs when a sentence is written in such a way that the subject receives, instead of completes, an action. Although it is not technically a grammatical error, using passive voice can often create wordier sentences and make a piece of writing less engaging. Whenever possible, it is best to revise passive voice to active voice. Here are some sentences using both passive and active voice.

> Passive voice: In the early twentieth century, hundreds of products using peanuts were developed by George Washington Carver.

Here, the subject, *George Washington Carver,* is receiving the action of the verb, *developed.*

> Active voice: In the early twentieth century, George Washington Carver developed hundreds of products using peanuts.

Using the same subject and verb, the subject, still *George Washington Carver,* is now the one developing the products.

> Passive voice: Supreme Court Justice Sonia Sotomayor was appointed to her position in 2009 by President Barack Obama.

The sentence above is wordy and awkward. The subject of the sentence is President Barack Obama, but it is difficult to tell since his name does not appear until the end of the sentence. Here is the same sentence again, this time using active voice.

> Active voice: In 2009, President Barack Obama appointed Justice Sonia Sotomayor to the Supreme Court.

Notice that by changing the sentence to active voice, fewer words were needed to express the same idea.

Pedantry

Pedantry in writing is an over and undue display of learning—one that can easily push readers away from the intended message. Pedantic writing often includes the unnecessary use of large words or the inclusion of minute details for the purpose of flaunting knowledge. For example, student writers commonly become pedantic in their academic writing because they feel the need to sound more "collegiate." The problem, however, is that students often do not know the meanings, or do not fully understand the meanings, of the words they are using.

Another reason student writers may appear pedantic in their writing is because they are trying to vary their words. However, this also can be problematic if the substitutions are simply pulled from a thesaurus, especially if the substitutions sound awkward and out of place. To avoid pedantic writing, it is best for a writer to use language that matches their own experience.

Below is an example of a pedantic sentence and a possible revision.

> Pedantic: The oppressive nature of the regime led to the despotic, tyrannical, and superincumbent exploitation of the local inhabitants.
>
> Revised: The oppressive nature of the regime led to the maltreatment of the local community.

The words *despotic, tyrannical,* and *superincumbent* all mean the same thing as *oppressive.* Thus, including all those words in one sentence is unnecessary. It is also okay to use more straightforward language, such as *maltreatment* instead of *exploitation,* to make a point.

Tone

Tone is an author's perspective on a subject. It can be identified through an author's choice of words or the way in which they engage with a topic. Furthermore, tone will, and should, vary depending on the writer's audience and purpose. For example, writing a text message to a friend will likely be casual in tone, and its purpose is commonly friendly and personal. However, writing an essay for a college class or perhaps for an industry journal means the tone will be more formal, and its purpose may be to inform, argue, or explain. In academic or professional writing, elements such as slang, contractions, and clichés should be avoided. An author who presents the wrong tone can inadvertently bury the primary message and derail their purpose. The wrong tone can also damage the author's relationship with the audience. Here are some examples of inappropriate and appropriate tones.

> Inappropriate: To identify the best solution for the city's traffic problems, the council should hold a public meeting and get the town's spin on things.

To get someone's *spin on things* means to get their thoughts on a subject. However, the phrase *spin on things* is slang. See below for a revision that eliminates the use of slang.

> Appropriate: To identify the best solution for the city's traffic problems, the council should hold a public meeting and hear citizens' concerns and ideas.

Notice that the phrase *get the town's spin on things* has been replaced by *hear citizens' concerns and ideas*. This replacement is both more formal and clearer in terms of its meaning.

Mood

Mood is the atmosphere a writer creates through their tone and word choice. Setting can also impact mood in a piece of writing, but depending on the text, there may or may not be a setting involved.

Take a look at this excerpt from Edgar Allan Poe's poem "The Raven."

> Ah, distinctly I remember it was in the bleak December;
> And each separate dying ember wrought its ghost upon the floor.
> Eagerly I wished the morrow;—vainly I had sought to borrow
> From my books surcease of sorrow—sorrow for the lost Lenore
> For the rare and radiant maiden whom the angels name Lenore

Here, the unnamed narrator is lamenting the loss of a loved one. In doing so, Poe carefully selects language that both reflects the setting and the narrator's mood. Words such as *bleak* and *sorrow*, for example, both describe an atmosphere filled with darkness and despair. The narrator also mentions that it is December, the depth of winter, when it is generally cold and the days are shorter. Thus, from this excerpt, we can identify the mood as being one of sadness and discomfort.

The purpose of mood in writing is to invoke emotions or appeal to a reader's senses. The intention here is to create a connection with the audience. Invoking an inappropriate mood for the subject being discussed can be detrimental to an author's meaning and credibility.

Mood can also be conveyed through the use of verbs. For example, in a research essay, an author generally seeks to portray facts to their readers. See the example below.

> Hurricane season *is* June 1 through November 30 each year.

The use of the word *is* presents a mood that is researched and non-negotiable. However, if an author wanted to propose a solution to a significant problem, the mood might appear as more urgent. Here is an example.

> All candidates *must* consider whether their platform on health care is truly inclusive.

The verb *must* reflects an imperative need for action, creating a mood that is critical and demanding. Here, the author specifically identifies a need (inclusive health care) and what they believe candidates must do to address it. The choice of verb clarifies the author's perspective on the subject.

Consistency

Having **consistency** means ensuring that the author uses certain elements the same way throughout the text. These elements generally include capitalization, numbers, and hyphenation. Consistently using these types of elements in the same way throughout a text increases a message's clarity as well as the author's credibility because purposeful attention is paid to detail.

Capitalization

Words that begin sentences and words that are proper nouns should always be **capitalized**. Common nouns, on the other hand, do not need to be capitalized; they refer to non-specific people, places, things, or ideas, such as *cat*, *house*, and *waterfall*. If we were to name a particular waterfall, however, it would be a proper noun and require capitalization (*Angel Falls*). Being consistent with capitalization means ensuring that the same words are always appropriately capitalized throughout a piece of writing. In short, grammatically correct and consistent capitalization clarifies for readers whether a person, place, thing, or idea is specific or generic. Here is an example.

> Mother appeared disheveled and confused. Her vehicle was located 12 miles from where she was found. When I arrived, the only thing mother could say was, "I don't know what happened." Doctors say she had suffered short-term amnesia.

In the example above, the first instance of the word *Mother* appears as the opening word of the first sentence. Because it begins a sentence, it should be capitalized. However, the narrator is also naming the mother in the sentence as their own parent. This makes this same instance of the word *mother* a proper noun, thus again requiring capitalization. In the third sentence, we see the word *mother* used once more. Here, too, the narrator is calling their parent by name: *Mother*. However, the second instance of the word is not capitalized. This is inconsistent because, in both cases, the word is referring to a specific, named person.

Numbers

The general rules for including numbers in writing are as follows. Note that this is typical for humanities writing; science and math writing have different rules. Numbers one through ninety-nine should be written out, and numbers higher than 100 should be written as numerals. For example, *seven thousand six hundred three* should be written as *7,603*. In addition, decimals should also be written out as numerals. For instance, *three point five percent* should be written as *3.5%*. Consecutive numbers should appear as numerical and alphabetical, such as *The camp divided children into groups of six 12-year-old campers per cabin.* If a number begins a sentence, it should be written out: *Twenty-four houses were slated to be built in one week.*

Hyphenation

Hyphens are punctuation marks that join two or more words together. There are several types of hyphenated words. Here are some examples of compound adjectives, compound words that modify a noun.

- Chocolate-covered
- Good-hearted
- Well-known

Hyphens are also used in writing ages, such as the following:

- Four-year-old son
- Six-year-old child
- Nine-year-old daughter

Hyphenate numbers when writing them out, usually from twenty-one to ninety-nine.

- Thirty-five
- Fifty-two
- Seventy-eight

Words that are combined to form a single word, such as *redhead* or *motorbike*, should not be hyphenated. These are known as *closed compounds*.

Expression of Ideas

Organization

Another significant area in a piece of writing that may need to be revised is organization. **Organization** refers to the way in which ideas are presented to readers, both via sentences and paragraphs. Proper organization is important because it ensures accurate interpretation of content by readers. If content is not logically organized, readers may become confused, misinterpret the author's intentions, and/or lose interest altogether. Below are discussions of specific strategies writers can employ to establish clear organization in their writing.

Transitions

Transitions are words or phrases that move ideas along and identify relationships between ideas. Furthermore, transitions tell the reader how to engage with the ideas a writer is presenting. Think of them as cues. If the writer is about to offer an opposition to an idea previously expressed, a transition word or phrase can help readers prepare for consideration of this new idea. Or, if a writer is adding new information to an idea, they may use a transition that tells readers that more information is going to be presented. Here is a list of common transitions used by writers. Note that there are many, many others.

Relationship Between Ideas	Transition Word or Phrase
Adding new information	In addition; also; furthermore
Cause and effect	As a result; consequently; therefore
Comparison	Similarly; likewise
Contrast	In opposition; however; in contrast
Conclusion	Thus; to summarize; finally
Illustrative	For example; to illustrate; for instance
Time or location	In the meantime; while; beyond

Transitions are critical to a text's flow, rationality, and clarity. If a writer seems to be missing a transition, look to see if there is a relationship between two or more ideas that may need clarification. Or, if sentences appear choppy (i.e., there is no flow), try and identify a transition that may help connect the author's ideas. Here is an example. In the revision, take note of the underlined words and phrases.

> Choppy: I went for a walk. I ended up at the park and sat down on the grass. I met a lost dog. She was very sweet; the owner came and retrieved her. I decided to research pet adoption.
>
> Revised: This morning, I decided to go for a walk. I ended up at the park and sat down on the grass. While relaxing, I met a lost dog. She was very sweet, and after a few minutes, the owner came and retrieved her. When I returned home that afternoon, I decided to research pet adoption.

The difference between the two examples is the addition of transitions. In the first, the writing sounds almost robotic. One idea is presented after another, but there is no continuity and no obvious connection between thoughts. In the revision, notice the addition of the underlined transitional words and phrases. All four are transitions that locate the author in time and place. Because of these transitions, we know when they went for a walk, when the dog greeted the author, when the owner showed up, and at which point they chose to look into pet adoption. Furthermore, the revised example simply reads more smoothly because of the addition of transitions.

Logical Sequence

In our discussion on transitions, the word *flow* was used more than once. When a piece of writing flows, it follows a logical sequence. In other words, the organization of sentences and ideas makes sense. If a writer does not employ a logical sequence of ideas, their writing may appear to be out of order and confusing. Three common ways a writer can create flow in their writing are cause and effect, chronological order, and comparison and contrast.

Cause and Effect

Cause-and-effect writing describes the reasons behind results or the results of certain actions or behaviors. Using cause-and-effect tactics can help a writer make connections between ideas by first discussing what happened or why something happened and then discussing any corresponding consequences or explanations. If done clearly and concisely, readers should easily be able to follow an author's thought patterns and come to the author's intended conclusions (in this case, what happened and why). Cause-and-effect writing generally includes transitions such as *consequently* or *as a result*, just to name a couple. If an author is attempting to employ cause-and-effect strategies but the connections are unclear, the writing may be in need of revision. Transitions may be needed, or ideas may need to be reorganized to make the causes and effects clearer to an audience. Ultimately, if the purpose is to express the how and why of something, clear cause-and-effect writing can help an author maintain flow.

Chronological Order

When something is written in **chronological order**, it is presented in a way that tells readers the order in which events occurred. Common transition words used in chronological writing are *first*, *second*, *third*, *next*, and *last*. Chronological writing may be used to describe at which point actions or experiences happened. At times, chronological order may also be referred to as **linear order**. To be **linear** means to appear in a straight line, with one thing happening or being placed after another. The proper use of chronological order can create a straightforward flow throughout a text. However, if a writer's chronological order is not linear, it may confuse readers, leading them to wonder when the events being discussed really took place. Again, like cause and effect, revision of chronological order may require the addition or reevaluation of transitions or organization.

Comparison and Contrast

To compare means to identify similarities and differences between two or more things, whereas to contrast means to find differences. In a **comparison and contrast text**, you might see transitions such as *in contrast*, *similarly*, or *in comparison*. For comparison and contrast to make sense, the text needs to be organized clearly, either by the elements being compared and contrasted or by individual points. For example, an author could discuss element A and all of the points they wish to make about that element, followed by a discussion of element B and all of its corresponding points. Or, an author may discuss each individual point as they relate to both elements A and B and then move on to the next point. However, if the writer jumps around from idea to idea without clear direction or connections, the similarities and differences may become lost to readers. In this case, there is likely an issue with organization, which can cause a hinderance in a text's flow.

Focus

Focus is critical to an author's work because it is the centerpiece upon which all other ideas in the text depend. Without focus, writing can quickly become confusing and frustrating for an audience to read. The purpose of an author's main idea is to control the direction of the content. Think about it like a road trip. If you are taking a long trip but do not have a route in mind, or even a destination, you can easily become lost. The same idea applies to reading and writing. As a reader, an author's focus allows you to identify the argument or dominant impression and follow it through to the end. A good author will also clearly connect subpoints to main ideas. If a piece of writing does not seem to have a main idea, or if the main idea is not clearly supported in the text, the writing may be in need of revision.

Introductions and Conclusions

Introductions and conclusions are like bookends: Without a solid one on either side of the books, the books may fall. The same can be said for introductions and conclusions in writing. A good introduction draws readers in and makes them want to keep reading. The author may do this with a shocking statistic, an interesting anecdote (real or hypothetical), or even an intriguing quotation. The introduction should also present readers with the text's main idea. A conclusion, on the other hand, should wrap up the writer's ideas but not restate them exactly the same way. A brief refresher of ideas may be helpful, but the writer should ultimately leave the reader with something to think about or a suggested action to take. However, a writer should avoid starting a new topic in the conclusion; it is important to remain on point to the very end.

Writing that lacks an interesting introduction can make readers want to stop reading before they have made it very far through the text. In addition, an introduction that does not introduce the text's main idea may also push readers away; a reader should not have to read very far into the text to find the main idea. If readers have to search for the main idea, the introduction may need work. Likewise, a conclusion that is repetitive or leaves a reader questioning the purpose of what they just read is indicative of writing that requires revision.

Supporting Ideas

As noted earlier, a piece of writing should have a clear focus. However, a main idea cannot drive a piece of writing on its own. It needs supporting ideas to help it along. If, for example, an author is making an argument that talking on handheld cellphones while driving should be banned in all fifty U.S. states, they will need to also provide supporting ideas that prove the validity of the point. These supporting ideas may be in the form of statistics, anecdotes, and expert testimony, all of which must support the need for a nationwide cellphone ban. A well-rounded argument includes a mix of all of these.

Writing that lacks supporting ideas for its main idea will need to be revised. For instance, if the author who is arguing for a nationwide cellphone ban while driving does not offer reasons or evidence to back up their claim, the main idea simply becomes an opinion or an unsubstantiated statement. Neither is strong enough to stand on its own, so the argument becomes null. A reader may also become confused as to why the writer is making such a claim, especially when no supporting points are provided. Other times, supporting ideas may need to be developed, or they may appear in an awkward place in the text. In these cases, a writer would want to revise the content by adding more support or revise the organization by moving the supporting ideas to a more accurate location.

Standard English Conventions

Sentence Structure

The **structure of a sentence** refers to the way in which a sentence is presented to readers. Structure includes the placement of words and phrases, the proper use of grammatical elements, and the level of interest and comprehension a sentence carries (i.e., its flow). There are four basic structures a sentence can follow, but since these have already been discussed, we will briefly list them here and provide an example for each.

Simple

A **simple sentence** has one subject and one verb with a single clause.

> Dr. Davison opened a family medical practice on Commerce Boulevard.

Subject: *Dr. Davison*; verb: *opened*; clause: *a family medical practice on Commerce Boulevard*.

Compound

A **compound sentence** includes two independent clauses, each with its own subject, verb, and clause. Sometimes, these sentences are separated by a comma and a coordinating conjunction or by a semicolon and a subordinating conjunction. Here are some examples.

> Dr. Davison opened a family medical practice on Commerce Boulevard, and he is accepting patients without insurance.
>
> Dr. Davison opened a family medical practice on Commerce Boulevard; furthermore, patients without insurance are being accepted.

In each compound sentence example, there are two independent clauses: *Dr. Davison opened a family medical practice on Commerce Boulevard* and *he is accepting patients without insurance* or *patients without insurance are being accepted*. In the first example, the clauses are separated with a comma and the coordinating conjunction *and*. In the second, they are separated by a semicolon and the subordinating conjunction *furthermore*. Also, in each example, there are two subjects: *Dr. Davison* and *patients*. In addition, each compound sentence has two verbs: *opened* and *is accepting/are being accepted*.

Complex

A **complex sentence** is a sentence that includes a dependent clause and an independent clause. Here is an example.

> Even though a patient may not have insurance, he or she may still make an appointment at Dr. Davison's new family medical practice on Commerce Boulevard.

The dependent clause in this example is *Even though a patient may not have insurance*. It is offset by a comma and followed by the independent clause *he or she may still make an appointment at Dr. Davison's new family medical practice on Commerce Boulevard*.

Compound-Complex

A **compound-complex sentence** is a sentence that includes both a dependent and two or more independent clauses. Here is an example.

> Because of his new policy, Dr. Davison is now accepting patients without insurance; his new practice is located on Commerce Boulevard.

The dependent clause is *because of his new policy*, which is offset by a comma and two independent clauses, *Dr. Davison is now accepting patients without insurance* and *his new practice is located on Commerce Boulevard*. The two independent clauses are separated by the semicolon.

The purpose of different sentence structures is to add variety to an author's writing. This kind of variety can break up potentially robotic and choppy sentences. Imagine if a passage read as follows:

> Simon bought a motorcycle. He had always wanted one. He drove it everywhere. He took it to the beach. He took it to the mountains.

There is nothing grammatically wrong with this paragraph. However, it lacks flow. Three of the sentences, for example, start with the same pronoun: *he*. As a result, the passage sounds mechanical. Here is a revision, using the sentence structures discussed above.

> Simon bought a motorcycle, something he had always wanted. He drove it everywhere. His first trip was to the beach, but his ideal trip was to the mountains.

Notice how the first sentence in the revised example is now a complex sentence with an independent clause, *Simon bought a motorcycle*, and a dependent clause, *something he had always wanted*. The second sentence, *he drove it everywhere*, remains a simple sentence, whereas the last sentences have been combined to form a compound sentence: *his first trip was to the beach, but his ideal trip was to the mountains.* The mixture of sentence structures creates a more interesting passage for an audience to read, which means they are more likely to remain engaged.

Usage

Usage refers to the conventional ways in which language is used, in this case, in the English language. These conventions are generally agreed upon by experts and are taught at all levels of academia.

The eight parts of speech are nouns, pronouns, verbs, adverbs, adjectives, preposition, conjunctions, and interjections. Sometimes, a ninth one, called a determiner, is included. Each one serves its own purpose in a sentence. In revision, it is important to identify whether parts of speech are appropriately being used and placed in a sentence.

Noun

A **noun** is a person, place, thing, or idea. Nouns can be common, such as *cat*, *house*, and *mitten*. Common nouns refer to generic people, places, things, or ideas. Nouns can also be proper, meaning they are specific names of people, places, things, or ideas. Examples of proper nouns are *Lake Michigan*, *Saint Michael's Catholic High School,* and *Stephanie*. Proper nouns must be capitalized, as should any noun that begins a sentence. Furthermore, a noun can be abstract, such as *happiness*, or concrete, such as *bookshelf*. Finally, nouns can be collective, such as *classes* or *storefronts*. Whichever form a noun takes, its purpose is to act as the subject of a sentence (the person or thing doing the action of the verb) or as the object of a sentence, such as the object of a preposition or a verb (the person or thing receiving the action of the sentence).

> Nouns as subjects: The <u>city of Baltimore</u> received an unprecedent amount of rain in only two days' time.

The *city of Baltimore* is receiving the rain; thus, it is completing the action of the sentence.

> Nouns as objects of verbs: Rowan could have sworn she had left her <u>keys</u> on her <u>kitchen counter</u>.

Although *Rowan* is a noun and is completing the action of the sentence, here we are focusing on the nouns, *keys* and *kitchen counter*. The noun *keys* is receiving the action of the verb *left*, and the noun *kitchen counter* is being located by the preposition *on*.

Pronouns

Pronouns are words that can act in place of nouns. Typically, a pronoun may be used to lessen the repetitive use of a noun or as an antecedent. An **antecedent** is a word that refers to the noun of the sentence. Furthermore, pronouns can take six different forms: personal, possessive, reciprocal, reflexive, demonstrative, and relative. A pronoun may also be first person, second person, or third person.

Personal pronouns refer to a specific person or thing. If we say, for example, *she went for an evening run*, the pronoun is *she*, which refers to a specific person in the sentence.

Possessive pronouns indicate ownership, as in *their car became stuck in the thick mud*. The pronoun *their* reflects ownership of the car.

Reciprocal pronouns act as antecedents because they refer to nouns in a sentence; furthermore, reciprocal pronouns indicate a mutual relationship between the nouns, as in *Teachers and graduates excitedly greeted one another in the auditorium.* The two nouns are *teachers* and *graduates*, and the reciprocal pronoun is *one another*.

Reflexive pronouns end in *-self* or *-selves* and are used when the subject of the sentence is the same as the object of the sentence, as in *The cat made herself comfortable on her owner's freshly laundered clothes. The cat* is the subject of this sentence, and the pronoun *herself* also refers to the cat.

Demonstrative pronouns are represented by the words *this, that, those,* and *these*. These pronouns generally represent things (i.e., objects) rather than people. In addition, without context, demonstrative pronouns do not fulfill a clear meaning. For example, if we say *Cameron thought that was a good idea*, we will need to know what *that* idea is in order to understand why he thought it was a good one.

Relative pronouns connect nouns and pronouns to phrases and clauses. Common relative pronouns include *where, who, whom, whose,* and *which*. An example of a relative pronoun used in a sentence would be *The bus driver who drove recklessly had her license suspended.* The relative pronoun *who* connects the noun *bus driver* to the phrase *drove recklessly had her license suspended*.

One problem to look for is **vague pronoun references**. A pronoun should have a clear antecedent, which is the noun to which a pronoun is referring. Here is an example.

> The mountain looked peaceful from a distance, but it could be deadly to ill-prepared climbers.

The pronoun in this sentence is *it*, and the noun to which the pronoun refers is *mountain*. Thus, the pronoun's reference is clear to readers. However, if a writer uses a pronoun such as *it, this, that,* or *which* but does not provide a clear antecedent, readers may be left wondering to what or whom the pronoun is referring. See the following example.

> The mountain looked peaceful from a distance. For some climbers, however, it was a deadly mistake.

Here, the writer uses the pronoun *it* in the second sentence, but it is unclear what the *deadly mistake* is and why. To revise for clarity, a writer should add a clear antecedent, or a noun or noun phrase that clarifies the writer's reference point. Here is a revised example of the sentence above.

> The mountain looked peaceful from a distance. For some ill-prepared climbers, however, the trip up the mountain was a deadly mistake.

Now we know that the pronoun *it* is referring to *the trip up the mountain*. The author has removed the vague pronoun and replaced it with a noun phrase.

Another thing to look for is **pronoun agreement**. Because pronouns can be singular or plural, it is important that they agree in number with their antecedents. Here is an example.

> Josephina was excited to finally receive her college degree.

Because *Josephina* is singular, the pronoun, *her*, is also singular. However, if we changed the sentence to *The administrators were excited to do their part in the graduation*, we now have a plural antecedent, *administrators*, and a plural pronoun, *their*. If a pronoun does not agree in number with its antecedent, we have an error.

Verbs

A **verb** can express three things: a physical action (run, jump, hide), a cognitive action (think, muse, consider), or a state of being (am, to be).

> Physical action: The children ran eagerly toward the snow cone truck.
>
> Cognitive action: Sienna thought carefully about her college options.
>
> State of being: The moving truck is here.

Other types of verbs include transitive, intransitive, and auxiliary. A **transitive verb** completes an action on a direct object, such as in *The teacher gave each student his or her own planner.* The verb here is *gave,* and the direct object (the thing being given) is the *planner*. **Intransitive verbs** are verbs that do not act on anything, as in *My head hurts*. Instead, the subject of the sentence, *head*, is the part of speech completing the action. Finally, **auxiliary verbs**, sometimes called **helping verbs**, are paired with main verbs to express an action, such as in *Conference attendees will be given a schedule of events and a map of the conference center.* The auxiliary (helping) verbs in this sentence are *will be*. They are paired with the main verb *given* to indicate the action that is being completed.

Adverbs

Adverbs modify verbs, adjectives, or other adverbs. In other words, adverbs give readers more information about the word being modified.

Adverbs can be used to modify a verb: Maria carefully dusted her grandmother's baker's rack, which was cluttered with fragile china and figurines.

The adverb *carefully* tells readers how Maria dusted her grandmother's baker's rack.

Adverbs can be used to modify an adjective: The audience enjoyed the play's delightfully witty characters.

The adverb *delightfully* describes the adjective *witty*.

Adverbs can be used to modify an adverb: After the storm, the street was <u>almost</u> completely flooded.

The adverb *almost* modifies the adverb *completely*.

Adjectives

Like adverbs, the function of **adjectives** is to modify other words. However, unlike adverbs, adjectives modify nouns and pronouns. Here are some simple examples.

Modifying a noun: The <u>angry</u> customer paced back and forth while she waited for the manager to arrive.

The adjective *angry* describes the noun *customer*.

Modifying a pronoun: He was <u>exhausted</u> after an intensive workout.

The adjective *exhausted* describes the pronoun *he*.

Generally, adjectives appear directly next to the noun or pronoun they are modifying, such as *the bald man*. In some cases, however, an adjective must be placed in a specific location. If modifying an indefinite pronoun, for example, the adjective should appear *after* the pronoun. An **indefinite pronoun** is a pronoun that does not specify a particular thing or person, such as *someone, anyone,* and *something*. An example would be *Something ominous is approaching.* Furthermore, adjectives may use premodifiers to help express a complete meaning. These are similar to auxiliary verbs, but in this case, a **premodifier** is added to a main adjective to create a complete adjective phrase: *The lake cabin is as wonderful as I remembered it to be*. The premodifier *as* helps the main adjective *wonderful* create a complete description of the cabin.

Another important function of adjectives is to express comparatives and superlatives. These are the degrees to which an adjective can be compared to something else in the sentence. In its base form, an adjective is called positive. For example, in its positive or **base form**, the adjective *fast* would be just that—*fast*. In its **comparative form**, it would be written as *faster*. Yet, in the highest form of comparison, or **superlative form**, it would appear as *fastest*. Not all adjectives can be formatted in such a unified way, however. Some adjective forms are *irregular,* meaning the entire word or large portions of the word must change for it to take a comparative or superlative form. For example, take the word *many*. We cannot say *manier* or *maniest*; these are not words. Thus, the comparative form of *many* would be expressed as *more*, whereas its superlative form is *most*.

Prepositions

Prepositions are words in a sentence that locate people, places, or things. A preposition must always be followed by a noun or a pronoun. Common prepositions include *in, on, above, under, over, after,* and *before*, but there are many others. In short, prepositions tell readers where something or someone is or when something took place. Here are some examples.

> The boys found an abandoned hut <u>on</u> the small island.
>
> <u>After</u> class, Ramona often spent time <u>in</u> the library studying the material covered that day.
>
> My grandparents' ranch is just <u>over</u> that hill.

Sometimes, a preposition may be mistaken for being part of a verb phrase. For example, the word *to* is a preposition. Look at the following sentence: *We often drive to the next town for groceries.* The verb is *drive*, and although the word *to* is directly next to the verb, it is actually part of the prepositional phrase (the preposition plus its modifying words), *to the next town*.

Conjunctions

Conjunctions are words that join ideas together, whether they are words, phrases, or clauses. There are three major types of conjunctions: coordinating, subordinating, and correlating.

Coordinating Conjunctions

Coordinating conjunctions include *for, and, nor, but, or, yet*, and *so*. Think FANBOYS. It is an easy way to remember the list of coordinating conjunctions. The purpose of a coordinating conjunction is to connect two ideas of equal emphasis. Here are some examples.

> Rosalie expected it to rain today, but she did not expect the large hail and high winds.
>
> School is set to begin next week, so stores are capitalizing on the need for back-to-school supplies.
>
> Dr. Jenkins was growing weary of his travel, for he had come a long way in a short amount of time.
>
> Michael searched for the lost dog on the east side of the neighborhood, and Nico searched the west.
>
> Neither the sons nor the father want to sell their family home.
>
> I am not sad, yet I am crying.

Using coordinating conjunctions correctly means identifying the most accurate and effective conjunction to appropriately join ideas of equal importance. Furthermore, proper punctuation must be used when separating ideas with conjunctions. Generally, a sentence using a coordinating conjunction looks like this: independent clause + comma + coordinating conjunction + independent clause.

Subordinating Conjunctions

Subordinating conjunctions are words that join together dependent and independent clauses. Their purpose is to indicate contrast, cause and effect, or other relationships between ideas. Common examples include *however, although, since, while, until*, and *because*. Below are some examples of sentences with subordinating conjunctions.

> However concerned you are, her decision is her own.
>
> I drove from store to store until I found the item I needed.
>
> The festival was even more crowded than expected, although most people stayed only for a short while.
>
> Since you are driving in that direction, please drop this package off at the post office.
>
> While the band warmed up, the crew checked the sound equipment.

The general rule when using a subordinating conjunction is this: If a dependent clause with a subordinating conjunction begins a sentence, that same clause must be followed by a comma and then the corresponding independent clause.

Correlating Conjunctions
Correlating conjunctions are conjunctions that pair together to create meaning. These conjunctions include *neither/nor*, *not only/but also*, and *either/or*. Here are some examples.

> Neither rain nor snow can prevent the town from holding their annual holiday celebration.
>
> Not only do we expect more from our school board, but we also expect better.
>
> Either you or Melinda will need to deliver the opening address.

Because sentences with correlating conjunctions typically have two or more subjects, the main verb should take the form of the closest subject. For example, in the example *Neither the sons nor the father wants to sell their family home*, *father* is the closest subject to the verb *want*. Since *father* is singular, the verb must also be singular (*wants*).

Interjections
Interjections are words used by writers to express a sudden feeling. Words such as *ouch*, *yikes*, and *oops* are examples of interjections. Although there are no specific rules for using interjections, they should be appropriately punctuated to help them make sense. Here are some examples.

> Ouch! That pan is hot!
>
> I accidentally submitted my final essay into the wrong drop box (oops!), but thankfully, I was able to resubmit it later that day.

Verb Tense
Verbs have three possible tenses: past, present, and future. Furthermore, each tense has different forms a base verb can take. The purpose of verb tenses is to tell readers when an action has occurred, is occurring, or will occur.

Here are some tables with examples.

Simple Present	Simple Past	Simple Future
I run a mile each day.	Last week, I ran a mile each day.	Next week, I will run a mile each day.

Present Continuous	Past Continuous	Future Continuous
I am planning a trip to Las Vegas.	Last night, I was planning my trip to Las Vegas.	I will be planning my trip to Las Vegas soon.

Present Perfect	Past Perfect	Future Perfect
I have unpacked most of the boxes in my kitchen.	By the time my friends arrived to help, I had unpacked most of the boxes in my kitchen.	I will have unpacked most of the boxes in my kitchen by early this evening.

Present Perfect Continuous	Past Perfect Continuous	Future Perfect Continuous
I have been swimming since I was a baby.	I had been swimming long before I took an actual swim lesson.	I will have been competitively swimming for five years before joining the school swim team.

Sometimes, a writer may mistakenly shift tense, causing confusion for readers and an interruption in a writer's flow. For example, if a writer is using past tense and shifts to present tense in the same sentence, this is likely an error. Here is an example.

> The plane took off from New York City that morning and will land in Los Angeles that evening.

The sentence begins in past tense with *took off*, but it ends in future tense. This is confusing and sounds awkward. To revise, both verb tenses should remain in the same tense. In addition, verb tenses should remain the same throughout an entire passage unless the writer clearly indicates that a shift is about to occur.

Capitalization Rules

Most **capitalization rules** are ones you have likely already been taught and know well. These include the following:

- Capitalize the first word of each sentence. Here is an example.
    - Because of the impending hurricane, the governor urged residents to evacuate low-lying areas. However, many people decided to stay and ride out the storm.

- Capitalize proper nouns. These include names of people and places.
    - On our summer vacation, we visited Calgary and Vancouver.
    - Dr. McGregor's philosophy class filled up before I could register for it.

- Capitalize the first word of a direct quotation, even if that quotation appears in the middle of a sentence.
    - Neil responded, "We will arrive around 6 o'clock tomorrow evening."

- Capitalize the titles of publications.
    - I enjoyed the movie *Titanic*.
    - My sister is reading the book *Midnight in the Garden of Good and Evil*.

- Capitalize days, months, and holidays.
    - August is the hottest month of the year.
    - On Thursdays, the office orders catering for all employees.

However, some capitalization rules are not as well known. These include words that should *not* be capitalized.

- Do not capitalize words after a colon, unless the word (or words) is a proper noun or begins a complete sentence. Here are some examples.
    - Her goal was clear: to win the race and move on to the semifinals.
    - There is only one place on my travel bucket list: New Zealand.
- Do not capitalize seasons.
    - The <u>winter</u> here is harsh, but the <u>summer</u> is pleasant.
- Capitalize major time periods and events.
    - The Civil War took place between 1861 and 1865.
    - Many inventions related to food consumption took place during the American Industrial Revolution.
- Capitalize languages and nationalities.
    - Susan is English, but she speaks German fluently.

Parallel Construction

When words and phrases in a sentence are **parallel**, this means they are grammatically the same. In other words, they match in structure and form. Sentence structures that are not parallel can be awkward to read.

For example, if a writer lists a series of verbs, those verbs should each appear in the same form. Here is an example.

> My favorite hobbies include <u>swimming</u>, <u>biking</u>, and <u>birdwatching</u>.

Notice how each verb (*swimming*, *biking*, and *birdwatching*) is a gerund, meaning they each end in *-ing*. So, the list is parallel.

When using connecting words, such as *between, and, both, neither/nor, either/or,* and *not only/but also,* the phrases or clauses that appear on either side of the connecting word must also be parallel in structure. Here are some examples.

> The class examined the relationships *between* the nations involved in World War I *and* the nations involved in World War II.
>
> The college campus is *both* beautiful and historic.
>
> *Neither* the road leading to our neighborhood *nor* the road leading to our house flooded from the immense rainfall.
>
> It was a pleasant surprise to learn that *not only* had the new facility opened on time *but* that it had *also* passed all necessary inspections.

Subject-Verb Agreement

When **subjects and verbs agree**, this means they match in tense and number. Here are the basic rules surrounding subject-verb agreement.

Singular subjects require singular verbs, and plural subjects require plural verbs. See the examples below.

> The stray cat visits the elderly woman's home each morning and each evening.
>
> Stray cats are often seen visiting the elderly woman's home in the mornings and evenings.

When using *either/or* and *neither/nor,* the verb takes the form of the closest subject.

> Either Joan or her sisters need to put their parents' house up for sale.
>
> Neither the girls nor their mother is picking up Uncle Stu from the airport.

Compound subjects require plural verbs.

> Father and Joseph plan to arrive early to the county fair.

Homonyms and Homophones

Homonyms
Homonyms are words that have the same spelling but a different meaning. Examples are words such as *refuse, project*, and *fair. Refuse*, for example, can mean a person's decision to not do something, but it can also mean garbage. Similarly, the word *project* can mean an activity or assignment and also mean to look into the future, and the word *fair* can represent a carnival, but it can also indicate equality and justice. These are only a few of the many look-alike/sound-alike words in the English language.

Homophones
Homophones are words that sound alike but are spelled differently and have different meanings. Homophones are one of the most confused types of words in the English language. Common examples

of misused homophones are *accept* (to receive something) and *except* (to exclude something or someone); *affect* (to impact something or someone) and *effect* (the result of something); *there* (a location), *their* (showing plural possession), and *they're* (a contraction meaning "they are"); and *to* (a preposition indication location or status), *two* (a number), and *too* (meaning also). Other examples are *your* and *you're, who's* and *whose,* and *its* and *it's.*

Double Negatives

A **double negative** is a grammatical error that occurs when a writer uses two negative words in a sentence. These negative words, in turn, cancel one another out. Here are some examples and revisions.

> Double negative: Scientists were not unconvinced by the new studies that had been released.
>
> Revision: Scientists were convinced by the new studies that had been released.
>
> Double negative: We didn't see nobody else on the train.
>
> Revision: We didn't see anyone else on the train.

Split Infinitives

An **infinitive** is the preposition *to* plus the base form of a verb, such as *to run* or *to see*. A **split infinitive** is a grammatical error that occurs when *to* is separated from the base verb. Generally, the split is created by an adverb or adjective. Note that not everyone agrees that a split infinitive is an error; to some, it is simply stylistic.

Here is an example of an infinitive and a split infinitive.

> Infinitive: We raced quickly to see the fireworks before they were over.
>
> Split infinitive: We raced to quickly see the fireworks before they were over.

Sentence Structure Errors

Common **sentence structure errors** include fragments, run-ons, and dangling and misplaced modifiers.

Fragments

A complete and grammatically correct sentence includes a subject, a verb, and a complete predicate. A subject is the thing or person doing the action in the sentence, whereas the verb tells the reader what action is taking place. A predicate is the verb plus any modifying words. Together, these elements form a complete and meaningful sentence. Here is a simple example.

> The kickboxing coach was running late for his class.

Our simple subject here is *coach*, and our complete subject is *kickboxing coach*. The kickboxing coach is the person doing the action of the verb, which, in this case, is *running*. The complete predicate is *running late for his class*. If we put the subject, verb, and complete predicate together, the sentence makes sense. A reader can understand the writer's point.

However, if a sentence is missing a subject, verb, or a complete predicate, it may be a fragment. A **fragment** is a group of words that is missing a critical element that would help it form a meaningful thought. Here are some examples.

> Fragment with a Missing Subject: Missing class Thursday because of an illness.

The example above does not tell us who is missing class on Thursday. It simply gives us a verb, *missing*, and a predicate, *on Thursday because of an illness*. Because we are missing important information, this example is a fragment.

> Fragment with a Missing Subject and Verb: On the desolate highway.

In this example, we do not know who or what is on the desolate highway or what they or it is doing there. If there was a verb, the predicate might be *on the desolate highway*, but without a verb, we do not know for sure. Thus, this group of words does not form a coherent thought. It, too, is a fragment.

Run-On Sentence

Run-on sentences are sentences that lack punctuation. Generally, a run-on includes more than one independent clause, or in other words, two sentences incorrectly fused together. Here is an example.

> Biologists were concerned about the troubling new behavior of the sharks they worked diligently to try to understand the underlying causes.

There are two independent clauses in the example above: *Biologists were concerned about the troubling new behavior of the sharks* and *they worked diligently to try to understand the underlying causes*. However, the two clauses are not separated by any punctuation, which makes the sentence confusing to read. To fix this error, the two clauses should be separated by a comma and a coordinating conjunction or by a semicolon. Another option is to simply separate the clauses with a period. Here are three possible revisions. Note the underlined additions.

> Biologists were concerned about the troubling new behavior of the sharks, so they worked diligently to try to understand the underlying causes.
>
> Biologists were concerned about the troubling new behavior of the sharks; therefore, they worked diligently to try to understand the underlying causes.
>
> Biologists were concerned about the troubling new behavior of the sharks. They worked diligently to try and understand the underlying causes.

As long as correct punctuation separates ideas at the appropriate points in a sentence, a writer can avoid run-on sentences.

Dangling and Misplaced Modifiers

Modifiers are words or phrases that add information to another word or phrase in the same sentence. Modifiers can take the form of gerunds or gerund phrases (verbs ending in *-ing*), prepositions or prepositional phrases, or adverbs and adjectives or adverb and adjective phrases. As a general rule, the modifier should be placed directly next to the word or group of words being modified. Here is an example.

> Running late, Ronaldo made it to the bus stop just in time.

The modifier in the sentence above is *running late*, and the word being modified is *Ronaldo*. However, if a modifier is placed somewhere other than next to the word it is modifying, this can create an error known as a **misplaced modifier**. See the example below.

> Peering through the window, the identity of the visitor slowly came into focus for Regina.

In this example, the modifying phrase is *peering through the window*. However, it has been placed next to *the identity of the visitor*. Because of this, the writer seems to indicate that the identity of the visitor is peering through the window, an action that is not possible.

Alternatively, if the person or thing being modified is not included in the sentence, this is an error called a **dangling modifier**. Here is an example.

> Growing up, the closest grocery store was 10 miles from our home.

The modifier in this case is *growing up*, but as written, it sounds like the grocery store is growing up. The subject being modified is missing from the sentence. Here is a revision.

> When I was growing up, the closest grocery store was 10 miles from our home.

Now, the modifier *growing up* is modifying the pronoun *I*. This revision allows readers to follow a writer's thoughts more clearly.

Punctuation

Punctuation refers to the marks used to separate sentences and elements of sentences. The most commonly used punctuation marks, besides periods, are commas, apostrophes, and semicolons. Here are the rules for each as well as common errors to avoid or revise.

Commas
The following are reasons to use commas in writing.

Before a coordinating conjunction and a second independent clause
This is the structure of a compound sentence. Here is an example: *The parking lot at the beach was full, but there were still several vacant spaces on the sand.* The first independent clause, *the parking lot at the beach was full*, and the second independent clause, *there were still several vacant spaces on the sand*, are separated by a comma and a coordinating conjunction.

To separate items in a list
When a writer includes a list, each element of that list should be separated by a comma. Here is an example: *Tomorrow we will clean the bathrooms, mop the kitchen floor, and vacuum the bedrooms.* Here, the writer lists three actions: *clean the bathrooms, mop the kitchen floor,* and *vacuum the bedrooms.* Each one is separated by a comma, including before the last item in the series. The comma

right before the last item in a series is called the *Oxford comma*. It should appear right before the word *and* or *or*.

After an introductory clause
An **introductory clause** introduces the main idea of a sentence. It is usually a dependent clause and must be offset by a comma. Here is an example: *To further support his point, the speaker shared several examples of his theory in practice.* The dependent clause here is *to further support his point*, which is then followed by a comma and the main clause: *the speaker shared several examples of his theory in practice.*

Commas in dates
In dates, a comma should be placed between the day of the month and the year: *September 4, 2018*. A comma should also be placed after the day of a week if it precedes a date: *Thursday, July 7, 1983.*

Commas before quotation marks
A comma should be placed before a quotation, right before the opening quotation mark. If a quotation does not end a sentence and a comma follows it, the comma should be placed inside the closing quotation mark: *John said, "I might be running late today," which was not unusual.*

Commas to offset non-restrictive clauses
A **non-restrictive clause** is a group of words that does not need to be in a sentence for the sentence to have a complete meaning. Usually, these words add information to a sentence but can be removed without confusing a reader. If used, a non-restrictive clause should be offset with commas: *The baseball team, along with their coaches, excitedly made their way to the state championships.* If the non-restrictive clause ends a sentence, a comma should be placed right before it: *Janice arrived early to the meeting, an unusual behavior for her.*

Common errors when using commas include comma splices (separating two sentences with a comma but no conjunction (*The sun is hot, put on some sunblock*), missing commas (*The sun is hot so put on some sunblock*), and unnecessary or misplaced commas (*The sun is hot so, put on some sunblock*).

Apostrophes
Apostrophes are used to indicate ownership (*Buck's new job*) or to create contractions (*they are* to *they're*). Because apostrophes only serve two main functions, there are very few rules about how to use them. For example, if a subject is singular, the apostrophe should be placed before the *s*: *The house's roof was leaking.* However, if a subject is plural, the apostrophe should appear after the *s*: *The teachers' professional development day was interrupted by a fire alarm.* The most common error a writer makes regarding apostrophes is unnecessary use or misplacement. For example, the sentence *The dog chased it's tail* is incorrect because in this case, the word *its* should be possessive (*its*) and not a contraction for *it is*.

Semicolons
Semicolons are used to separate independent clauses or, in some cases, items in a list. When used to separate two independent clauses, there should be a relationship between the two. Here is an example: *Sharona enjoys horror movies; however, Jace prefers comedies.* When used in a list, a semicolon may separate elements that also include commas. For example, if an author is listing cities and states, a semicolon can be used to create clarity. An example would look like this: *Our road trip took us through Oklahoma City, Oklahoma; Tucson, Arizona; and Sacramento, California.* A common error when using

semicolons is placing them between a dependent clause and an independent clause: *Because we were first to speak; we arrived at the event early*. Here, the clauses should be separated by a comma, not a semicolon.

Colons
Colons are used to introduce a list or modifying information. In other words, a colon directs readers to additional information in a sentence. Unlike a semicolon, the content that follows a colon does have to be an independent clause. Here are two examples.

> Introducing a list: When at the supermarket, please do not forget to pick up the following items: tomatoes, garlic, and olive oil.

Notice how the colon is placed directly after the independent clause *When at the supermarket, please do not forget to pick up the following items* and before the first item in the list.

> Adding information: Brit has only one goal this semester: to improve her understanding of geometry.

Again, the colon appears after the independent clause in the sentence and right before the dependent clause. However, this time, instead of a list, the writer provides more information about the subject's (Brit's) goal.

Hyphens
Hyphens are used to join words or parts of words. For example, a **compound modifier** is a combination of two words that work together to act as an adjective. These types of modifiers generally appear right before a noun. To be joined correctly, compound modifiers should be hyphenated. Here are some examples: *dog-friendly hotel*, *two-faced people*, and *well-known actor*. Other words that should be modified are compound nouns, such as *sister-in-law* or *seven-year-old*.

Dashes
There are two types of **dashes**: an em-dash and an en-dash. **Em-dashes** can be used to replace commas, parentheses, or colons. For example, when a writer includes non-essential information in a sentence but does not want to place this information in parentheses or offset it with a comma, an em-dash may be used. In addition, the em-dash may be used to offset this information in the middle or at the end of the sentence. Look at the following examples.

> Faye (who was already agitated) begrudgingly made her way to the staff meeting.
>
> Em-dash: Faye—who was already agitated—begrudgingly made her way to the staff meeting.
>
> Faye, who was already agitated, begrudgingly made her way to the staff meeting.

En-dashes, on the other hand, are used to show numerical durations, such as *1 p.m.–6 p.m.* or *1960–1980*. En-dashes can also be used to connect compound words that already contain hyphens, such as *pro-life–pro-choice*.

Quotation Marks

Quotation marks should be used to identify another person's words. These other words can be from another person's written or spoken works or dialogue that has been put into writing. Quotation marks should also be used around titles of short works, such as the titles of journal articles. More rules for using quotation marks are as follows:

A quotation requires that a quotation mark should appear on either side of the quotation: Officer Strickland stated that he had "arrived at the accident scene at the same time as the ambulance."

If the first word of a quotation begins a full sentence, capitalize it: The newspaper article reported, "All residents had been evacuated due to an approaching wildfire." If the quotation is grammatically part of the non-quoted words, capitalization is not necessary unless there is a proper noun: According to historians, playwright William Shakespeare "wore a gold hoop earring, a Bohemian look."

Commas, exclamation points, and periods should always be placed inside quotation marks, whereas dashes and semicolons should be placed outside the quotation marks.

> Shayna's mother called out to her, "It's time to eat!"
>
> The librarian told the patron that the study rooms were "booked until next Thursday."
>
> He had only one thing to say to her—"Make sure you keep your eye on the prize."
>
> Dad said, "We'll see you when you arrive"; unfortunately, the flight was delayed by three hours.

If a quotation appears within another quotation, single quotation marks should be placed around the inside quotation: Biologist Anthony Menendez noted that "with a larger focus on breeding disease-resistant plants, we can eliminate what farmers claim to be a 'serious worldwide food deficiency'."

Ellipses

Ellipses (three dots, with a space before and after) indicate an omission. This means the writer has intentionally left words out of a quotation. A writer may do this because the omitted words are irrelevant to the rest of the sentence or because they would cause awkwardness if left in the sentence. Here is an example.

> Charles Dickens once wrote that, "The world belongs to those ... with self-confidence and good humor."

However, if an ellipsis appears at the end of a sentence, a period should also be included.

Analysis in History/Social Studies and in Science

The PSAT Writing and Language test will also test your ability to read, comprehend, and analyze passages in history, social studies, and science. You will then be asked to make decisions about how to best revise these passages. Keep in mind that some passages may be paired with images, tables, graphs, or charts. It is important that you examine both the passage and corresponding image, table, graph, or chart and identify how each supports the other. Some of the questions you can expect are based on the relationship between passage and informational graphic. As an example, you may be asked to look at the sentence level of a passage and identify revisions for a misinterpretation represented in a corresponding informational graphic. You will NOT be tested on prior knowledge, so it is okay if you are not familiar with the content in one or more of the passages.

Practice Quiz

Questions 1–5 are based on the following passage:

Outdoor Safety Tips for Lightning Preparedness

The best defense is to avoid lightning. Here are some outdoor safety tips that can help you avoid (1) <u>to be</u> struck:

Be aware
Check the weather forecast before participating in outdoor activities. If the forecast calls for thunderstorms, postpone (2) <u>you're</u> trip or activity, or make sure adequate safe shelter is readily available.

Go indoors
Remember the phrase, "When thunder roars, go (3) <u>indoors.</u>" Find a safe, enclosed shelter when you hear thunder. Safe shelters include homes, offices, shopping centers, and hard-top vehicles with the windows rolled up.

Seek shelter immediately even if caught out in the open
(4) <u>If you are caught in an open area act quickly to find adequate shelter</u>. The most (5) <u>importent</u> action is to remove yourself from danger. Crouching or getting low to the ground can reduce your chances of being struck, but does not remove you from danger.

Excerpt adapted from the CDC Lightning Safety Tips Page, from
https://www.cdc.gov/disasters/lightning/safetytips.html

1. Choose the correct answer.
    a. No change
    b. being
    c. are
    d. to being

2. Choose the correct answer.
    a. No change
    b. you
    c. you are
    d. your

3. Choose the correct answer.
    a. No change
    b. indoors".
    c. indoors'.
    d. indoors.'

4. Choose the correct answer.
   a. No change
   b. If you are caught in an open area. Act quickly to find adequate shelter.
   c. If you are caught in an open area, act quickly to find adequate shelter.
   d. If you are caught in an open area; act quickly to find adequate shelter.

5. Choose the correct answer.
   a. No change
   b. importint
   c. important
   d. importunt

Answer Explanations

1. B: The best choice is *B*, being, which reads "tips that can help you avoid being struck." The verbs *are* as well as *being* are forms of the infinitive *to be*, but *are* and *to be* do not fit here, making Choices *A* and *C* incorrect. *To being*, Choice *D*, does not work in this sentence either; only *being*, which is a present participle that indicates a continuous action.

2. D: The correct answer is Choice *D*, your. Choice *A*, you're, is a contraction for *you are*, which is incorrect, as it's not a possessive for trip in the sentence. Choices *B* and *C*, you and you are, are also not possessives, so they are incorrect.

3. A: The choice is correct as-is. Punctuation goes inside the quotation mark. Choice *B* is incorrect because it shows the period outside of the quotation mark. Choices *C* and *D* are incorrect because they use a single quote instead of a double quote, which is not appropriate for quoting material in standard American English.

4. C: Choice *C* is the correct answer because it separates an independent and dependent clause with a comma. Not all clauses are always separated by a comma, but if the dependent clause comes first, as in this sentence, it will require a comma after. Choice *A* is incorrect because it is lacking the comma after the dependent clause. Choice *B* is incorrect because it creates a fragment with the first sentence. Choice *D* is incorrect because a semicolon should have two independent clauses on each side, and the first clause is dependent.

5. C: This is a spelling error question. The correct way to spell important is Choice *C*. The other answer choices are incorrect.

Math

Heart of Algebra

Solving a Linear Expression or Equation in One Variable

A **linear equation in one variable** x can be expressed in the form $ax + b = 0$, for real numbers a and b. Note that $a \neq 0$. For example, $4x - 5 = 19$ and $-3x + 2 = 0$ are linear equations in one variable. The solution of a linear equation is a value that, when plugged in for x, makes the equation true. In order to solve a linear equation, two principles can be used. The **addition principle** states that if $a = b$, then $a + c = b + c$. Basically, a value can be added to one side of an equals sign in an equation as long as that same value is added to the other side. The addition principle also works for subtraction because subtraction is equal to adding the opposite of a number. Therefore, if $a = b$, it is true that $a - c = b - c$. This principle could be used to solve the equation $x - 6 = 19$. The final step in solving an equation involves isolating the variable. Therefore, 6 can be added to both sides to obtain $x - 6 + 6 = 19 + 6$, or $x = 25$.

The **multiplication principle** can also be used to solve equations. This principle states that an equation remains valid if the same real number is multiplied times both sides of the equals sign. Therefore, if $a = b$, it is true that $ac = bc$. In a similar fashion, if $a = b$, $\frac{a}{c} = \frac{b}{c}$. Therefore, the same value can be divided off of both sides of the equals sign in an equation. The multiplication principle can be used to solve the linear equation $8x = 32$. Dividing both sides by 8 results in $x = 4$. Both the addition and multiplication principles can be used to solve the equation $5x + 6 = 41$. First, subtract 6 off of both sides to obtain $5x = 35$. Then, divide both sides by 5 to obtain the solution $x = 7$. A solution can be checked by plugging it into the original equation. If it results in a true statement, the solution is verified.

Linear equations can be utilized in real-world scenarios. For instance, if a taxi driver charges an initial fee of $4 to enter her car and an additional $2 per mile driven, the cost of a taxi ride in algebraic form is $4 + 2x$, where x is the total number of miles driven. If $16 was spent on a taxi ride, the number of miles driven could be found algebraically. First, set $16 equal to the cost expression to obtain:

$$16 = 4 + 2x$$

Then, subtract 4 off of both sides, resulting in $12 = 2x$. Dividing both sides by 2 results in $x = 6$. This solution shows that a 6-mile taxi ride cost $16.

Solving Linear Inequalities in One Variable

A **linear inequality in one variable** x can be expressed in a similar manner as a linear equation $ax + b = 0$, $a \neq 0$. However, the main difference is that the equals sign is replaced by one of the following inequality symbols: <, >, ≤, or ≥. The symbol < is read as "less than." The symbol > is read as "greater than." The symbol ≤ is read as "less than or equal to." The symbol ≥ is read as "greater than or equal to." Specific examples of linear inequalities in one variable are:

$$7x + 13 < 0$$

and

$$9x - 2 \geq 0$$

To solve an equality, the set of numbers that satisfies the mathematical statement must be found. The **solution set** of an inequality consists of the entire set of numbers that, when plugged into x, make the inequality true. To solve an inequality, the same properties that are used in solving equations can be applied. The addition principle can be applied to either add or subtract variable terms and constants so that all variable terms are on one side of the inequality symbol and all constant terms are on the other side of the inequality symbol. The multiplication principle can be used to either multiply both sides times a real number or divide both sides by a real number to clear any coefficients attached to the variable. This last step will result in an inequality that gives the solution set. The multiplication principle applied to inequalities changes when working with negative numbers. When multiplying or dividing by a negative number, change the direction of the inequality symbol. In other words, "flip the sign." The solution set can be expressed in interval notation and graphed on a number line.

Consider the linear inequality:

$$-7x - 16 > 2x + 11$$

In order to find the solution set, add 16 to both sides and subtract $2x$ off both sides to obtain $-9x > 27$. Note that all variable terms are on one side, and all constant terms are on the other. Then, divide both sides by –3, making sure to "flip" the inequality symbol. The solution set $x < -3$ is obtained. Therefore, any number less than –3 satisfies the original inequality when plugged in for x. This solution in interval notation is $(-\infty, -3)$. The parenthesis shows that –3 is not actually included in the solution set, but all values less than –3 are.

Linear inequalities can also be used in real-world settings. Phrases such as "at least" or "at most" indicate inequalities. For instance, using the same taxi example as before, if the taxi driver wanted to charge at least $30 for a ride, the inequality in algebraic form is:

$$4 + 2x \geq 30$$

Finding the solution set would result in the number of miles the driver would need to drive to ensure a $30 fare. First, subtract 4 off both sides, and then divide by 2 to obtain $x \geq 13$. Therefore, driving at least 13 miles would result in a fare of at least $30. Notice that if the driver wanted to charge more than $30 in this scenario, the inequality symbol would change to $>$ in both the inequality and the solution set.

Linear Functions that Model a Linear Relationship Between Two Quantities

A **linear function modeling a linear relationship between two quantities,** x **and** y, is of the form:

$$f(x) = mx + b$$

Because $f(x) = y$, its equivalent equation form is written as:

$$y = mx + b$$

The graph of a linear function is a line with **slope** m and $y-$intercept b. Each point on the line is an ordered pair $(x, f(x))$. There is a relationship between the two variables. Note that x gets plugged into the function to obtain $f(x)$ or y. Because the function is linear, the variable y depends linearly on the variable x. The slope is also known as the rate of change of the function. For every one unit increase in x, $f(x)$ increases by m if the slope is positive and decreases by m if the slope is negative. If the slope is positive, the graph increases from left to right. If the slope is negative, the graph decreases from left to right.

A common example of a linear function involves cost and revenue functions. A **cost function** is a function that determines the cost of producing x units of a product. Consider the cost function:

$$C(x) = 15x + 100$$

This function shows that there is an initial cost of $100 to produce a number of units and a cost of $15 per unit. For example, the cost of producing 5 units is:

$$C(5) = 15 \times 5 + 100 = \$175$$

A **revenue function** is a function that determines the revenue of selling x units of a product. Typically, a revenue function does not have a $y-$intercept. Consider the revenue function:

$$R(x) = 20x$$

This function shows that the revenue of selling an individual unit is $20 and the revenue of selling 5 units is:

$$R(5) = 20 \times 5 = \$100$$

Because **profit** is equal to revenue minus cost, the profit function for this example is:

$$P(x) = R(x) - C(x)$$

$$20x - (15x + 100) = 5x - 100$$

The **breakeven point** is found by setting the profit function equal to 0 and solving for x. Because $5x - 100 = 0$ implies $x = 20$, this company must sell 20 units to break even and more than 20 units to make a profit.

Solving Systems of Linear Inequalities in Two Variables

A **system of linear inequalities in two variables** consists of at least two inequalities and two variables x and y. For instance, an example of such a system can be seen here:

$$\begin{cases} 6x + 4y \leq 1 \\ 6x - y \leq 2 \end{cases}$$

Within this system, the x and y are related because the solution set of a system contains all ordered pairs (x, y) that satisfy both individual inequalities. The usual way a solution set of a system is represented is through its graph on the **Cartesian plane**. Typically, a solution set of a system of linear inequalities is a half-plane, or a smaller portion of the Cartesian plane.

In the real world, a system of inequalities can be used to find the best answer to a problem. For instance, let's say that someone works two jobs. The first job pays $10 an hour, and the second job pays $12 an hour. This same person has $200 of bills each week and can work no more than 18 hours each week. Let x represent the number of hours worked each week at the $10 an hour job, and let y represent the number of hours worked each week at the $12 an hour job. The following system would represent this scenario:

$$\begin{cases} x + y \leq 18 \\ 10x + 12y \geq 200 \end{cases}$$

The first inequality states that the total number of hours worked in both jobs must be at most 18 hours. The second inequality describes total weekly income. The expression $10x$ represents the total amount earned from the first job, the expression $12y$ represents the total amount earned from the second job, and the sum of both must be at least $200. The solution set of this system of inequalities would represent all the possible combinations of hours worked at either job that satisfy both inequalities.

Solving Systems of Two Linear Equations in Two Variables

A **system of two linear equations in two variables** x **and** y consists of two individual equations. An example can be seen here:

$$2x + 7 = 15$$
$$5x - 8 = 14$$

The two variables are related because the solution consists of any ordered pair (x, y) that satisfies both equations. There are three possibilities for solutions to a system of two equations in two variables. There can be one ordered pair solution, there can be no solution, or there can be infinitely many solutions represented by a single line.

Systems of linear equations exist in the real world. For instance, let's say that a school is holding a bake sale. Each brownie costs $1.25, and each cookie costs $.75. At the end of the sale, the school made $112.50 and sold 110 individual items, but did not keep track of the number of individual brownies and cookies sold. A system of linear equations could help the school figure this information out. If x represents the total number of brownies sold, and y represents the total number of cookies sold, the following system represents the scenario:

$$x + y = 110$$

$$1.25x + .75y = 112.50$$

The first equation states that the total sum of brownies and cupcakes was equal to 110. The second equation states that the total amount earned was $112.50 because the expression $1.25x$ represents the total amount earned from the brownies and the expression. $75y$ represents the total amount earned from the cookies.

A way to solve a system of equations is to graph the two equations. If the lines intersect at a single point, there is one solution. If the lines do not intersect at all, there is no solution, and the lines are parallel. If the lines lie on top of one another, there are infinitely many solutions. The following graph represents the solution of the bake sale example:

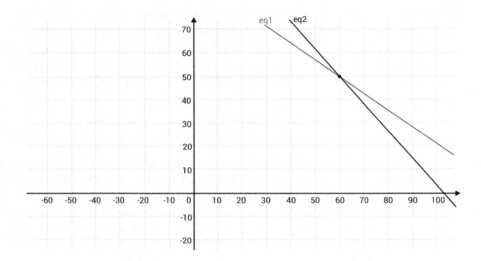

Notice that the lines intersect at the ordered pair (60, 50). Therefore, there were 60 brownies sold and 50 cookies sold.

Algebraically Solving Linear Equations in One Variable

Both linear equations and linear inequalities in one variable can be solved using the addition principle and the multiplication principle. The goal in solving an equation is to isolate the variable. The steps involve "moving" terms and coefficients away from the variable to the other side of the equals sign or inequality symbol. Basically, whatever operation exists within the expression, the opposite needs to be completed to isolate the variable. Also, if any fractions or decimals are in the equation or inequality, they can be cleared algebraically to make the solution process easier.

For instance, consider the equation:

$$\frac{9}{2}x + \frac{7}{10} = 5$$

The fractions can be cleared by multiplying every term within the equation times 10, which is the LCD of $\frac{9}{2}, \frac{7}{10}$, and $\frac{5}{1}$. This process results in:

$$45x + 7 = 50$$

Then, subtract 7 off of both sides to obtain $45x = 43$. Finally, divide both sides by 45 to obtain the solution $\frac{43}{45}$. This solution can be checked by plugging it into the original equation and ensuring a true mathematical statement.

The solution to an equation can be all real numbers or no solution. Consider the equation:

$$5x + 3 + 3x = 9x - x + 6$$

First, combine like terms to simplify either side. This process results in:

$$8x + 3 = 8x + 6$$

Next, subtract $8x$ from both sides. This last step results in $3 = 6$, which is a false statement. The original equation has no solution because there is no value for x that can ever make the solution true. No solution is represented in symbolic form as $\emptyset$. Secondly, consider the equation:

$$4(x + 5) = 4x + 20$$

To solve this equation, distribute the 4 to clear the parentheses. This step results in

$$4x + 20 = 4x + 20$$

Notice that both sides of the equals sign are the same. If $4x$ and 20 are subtracted from both sides of the equals sign, the result is $0 = 0$, which is always true. There are infinitely many solutions to this equation because any real number is a solution. All real numbers can be represented symbolically as $\mathbb{R}$ or in interval notation as $(-\infty, \infty)$.

Inequalities in one variable can be solved using similar algebraic techniques. Consider the inequality:

$$0.5x + 1.2 \leq 10$$

First, the decimals can be cleared if desired to ease the solution process. Because the decimals are in the tenth place, multiply every term in the equation times 10 to obtain:

$$5x + 12 \leq 100$$

Then, subtract 12 off of both sides to obtain $5x \leq 88$. Finally, divide both sides by 5 to obtain the solution set $x \leq \frac{88}{5}$. Therefore, any value less than or equal to $\frac{88}{5}$ satisfies this inequality. Recall that if an inequality needs to be multiplied times a negative number or divided by a negative number, the inequality symbol gets "flipped."

Algebraically Solving Systems of Two Linear Equations in Two Variables

There are two algebraic processes for solving systems of two linear equations in two variables. The first process is referred to as **substitution**, and it should be used on systems when one of the equations is already solved for either x or y or if solving for either would be a simple process. A system that has an equation in **slope intercept form**, $y = mx + b$ form, is best suited for substitution. For instance, consider the following system:

$$y = 5x - 6$$

$$3x + 2y = 10$$

The first equation is in slope intercept form and is solved for y. This equation is substituted in for y in the second equation to obtain $3x + 2(5x - 6) = 10$, a linear equation in one variable. This equation is then solved for x using the addition and multiplication principles. First, distribute the 2 and collect like terms, resulting in:

$$13x - 12 = 10$$

Then, add 12 to both sides and divide by 13 to obtain $x = \frac{22}{13}$. Because the solution of a system involves an ordered pair, the corresponding y-value must be found. Plugging $x = \frac{22}{13}$ into the first equation results in:

$$y = 5\left(\frac{22}{13}\right) - 6 = \frac{110}{13} - 6$$

$$\frac{110}{13} - \frac{78}{13} = \frac{32}{13}$$

The solution to the system is $\left(\frac{22}{13}, \frac{32}{13}\right)$. The substitution method could also be applied when one of the equations was originally solved for x.

The other algebraic method for solving systems of linear equations is referred to as the **addition or elimination method**. This process is better suited when both equations are in standard form $Ax + By = C$. The steps in this method involve multiplying one or both equations times real numbers that result in pairs of variables with opposite coefficients. Then, by adding equations together, those pairs cancel each other out, and an equation in one variable is obtained, which can be solved easily. Then, that solution value can be used to find the value of the other variable in the ordered pair solution. For example, a system that should use elimination to solve it is:

$$5x + 2y = 6$$
$$6x - 4y = 10$$

The substitution approach would become very complicated on this system because solving for any of the variables in either equation would result in many fractions. In order to use elimination, multiply the entire first equation times 2. This step results in:

$$10x + 4y = 12$$

Notice that now the coefficients on y are opposite. Adding the two equations together results in $16x = 22$. The solution to this equation is:

$$x = \frac{22}{16} = \frac{11}{8}$$

Plugging this back into either original equation results in $y = -\frac{7}{16}$.

A solution to a system of equations can be checked by plugging the entire ordered pair into both original equations. If the process results in a false statement, an error has been made. Also, if at any point throughout the solution process, a step results in a false statement, there is no solution to the system. Graphically, no solution is the case of parallel lines and no point of intersection. Also, if at any point throughout the solution process, a step results in a mathematical statement that is always true, there are infinitely many solutions. Graphically, this would be the case in which the lines lie on top of each other.

Interpreting the Variables and Constants in Expressions for Linear Functions

A linear function is of the form $f(x) = ax + b$, where a and b are real numbers. In this case, the x variable is the input and is known as the **independent variable**. It is true that $f(x)$ and y are interchangeable in function form and therefore, y is the output and is referred to as the **dependent variable**. The dependent variable depends on the independent variable. In a linear function, the output variable changes linearly with respect to the input variable. For example, for the following function:

$$f(x) = 6x + 5$$

$$f(1) = 6(1) + 5 = 11$$

$$f(2) = 6(2) + 5 = 17$$

$$f(3) = 6(3) + 5 = 23$$

Note that for every increase in 1 unit in the input variable, the output variable increases by 6. The graph of a linear function is a straight line. The rate of change of this particular function is 6, which is equal to the slope of the line. The linear function can also be expressed as $f(x) = mx + b$ where m is the slope and b is the y-intercept.

A real-world application of a linear function involves a **cost function**, which describes the cost associated with creating or manufacturing a certain number of items. For instance, let's say that the cost associated with baking x pizzas at a local pizza parlor is:

$$C(x) = 4x + 2$$

Therefore, for any number of pizzas, there is a flat cost of $2 and a cost of $4 per pizza. The $2 could be overhead costs, such as the cost of turning on the oven or wages of the pizza baker, which will occur no matter how many pizzas are made. The cost of 20 pizzas is found by plugging 20 into the function:

$$C(20) = 4(20) + 2 = 82$$

Therefore, it costs the pizza parlor $82 to make 20 pizzas. The rate of change, or slope, of the cost function is 4 and it is equal to the unit cost per pizza. The $y-$intercept of the function is 2, which is equal the flat cost of $2. Note that this quantity is found by plugging 0 into the function. It is technically the cost of making no pizzas.

Connections Between Algebraic and Graphical Representations

When a linear equation is in standard form $Ax + By = C$, it is hard to make sense of what the graph of the equation looks like from the coefficients alone. Ordered pairs can be found to graph the line. However, if the equation is in slope intercept form $y = mx + b$, the slope m and $y-$intercept b can be used to quickly create a graph. Therefore, it is helpful to place any equation in its slope-intercept form if a graph is desired. In other words, solve the equation for y.

Consider the standard form equation:

$$4x + 7y = 15$$

To place it in slope-intercept form, algebraic solution techniques can be used. Using the addition principle, subtract $4x$ off of both sides to obtain:

$$7y = -4x + 15$$

Then, using the multiplication principle, divide both sides by 7 to obtain:

$$y = -\frac{4}{7}x + \frac{15}{7}$$

The slope of this line is $-\frac{4}{7}$. Therefore, for every one unit increase along the x-axis, the y-value decreases by $-\frac{4}{7}$. The y-intercept of the line is $\frac{15}{7}$. Therefore, the point at which the line crosses the y-axis is the ordered pair $\left(0, \frac{15}{7}\right)$. The graph of the line can be seen here:

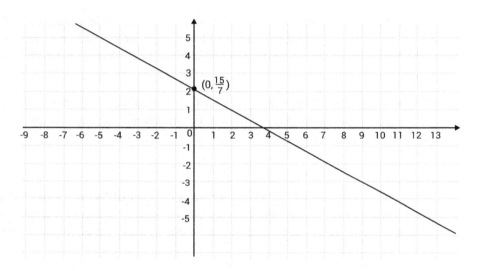

The line is decreasing from left to right, which is consistent with the negative slope. A line that increases from left to right has positive slope. The y-intercept lies above the origin, which is consistent with the fact that its value is positive. If the y-intercept was negative, the place where the line crosses the x-axis would lie below the origin.

Problem Solving and Data Analysis

Ratios, Rates, Proportional Relationships, and Scale Drawings

A **ratio** is equal to the quotient of two quantities. For instance, the ratio $\frac{3}{4}$, can be read as the ratio of 3 to 4, and can also be written as $3:4$ using colon notation. Ratios appear in everyday life. For instance, if a person spends on average 42 hours a week working and 51 hours a week sleeping, the ratio of hours spent working to sleeping each week is $\frac{42}{51}$ or $42:51$. In this example, both the numerator and the denominator compare the same type of measure, which is hours. If a ratio compares different types of measure, it is called a **rate**. Rates also appear in daily life. For example, an hourly wage of $10 an hour is a rate. Also, driving 65 miles per hour is a rate.

A common example of rates that appears in everyday life corresponds to unit costs, which are equal to the ratio of the price of an item to the number of units in that item. For instance, if a 64-ounce bottle of soda cost $2.59 and a 20-ounce bottle of soda cost $1.65, calculating the unit cost would show which item was a better deal. A better deal in this case would mean a lower cost per ounce. To find the unit cost, divide the cost by the number of ounces. The 64-ounce bottle of soda has a unit cost of $\frac{\$2.59}{64} \approx$ $0.04 per ounce. The 20-ounce bottle of soda has a unit cost of $\frac{\$1.65}{20} \approx \$0.08$ per ounce. Therefore, the larger bottle of soda is a better deal because it has a lower unit cost.

When two ratios are set equal, a **proportion** is formed. An example of a proportion is $\frac{3}{4} = \frac{6}{8}$. If one of the amounts in a proportion, either a numerator or a denominator, is an unknown, it can be solved for by using cross products. For instance, consider the proportion $\frac{4}{5} = \frac{11}{x}$. Cross multiplication results in equating the cross products as $4x = 55$.ND dividing by 4 results in the solution $x = \frac{55}{4}$.

Similar triangles are triangles in which the lengths of corresponding sides are proportional. Therefore, the lengths of corresponding sides also have the same ratio. This fact can be used to find missing side lengths in certain problems. Consider the following example of similar triangles:

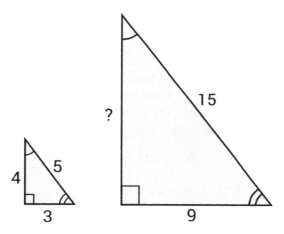

Because corresponding side lengths have the same ratio, the following proportion can be set up and solved to find the missing quantity:

$$\frac{15}{5} = \frac{?}{4}$$

Cross-multiplying and dividing by 5 results in a missing side length of 12.

Ratios and proportions can also be used to answer questions regarding scale drawings. A **scale drawing** is a drawing that either has been reduced or enlarged from its original size, given a specific scale. Maps are examples of scale drawings, and an example of a scale would be where an inch on the map represents a mile in actual real life. In this case, the original size has been reduced. For another example, let's say that in the classroom, a student is tasked to draw a scale drawing of her living room. If the scale factor of the drawing is to be one-twelfth of the original size, she would have to scale down the size of her room in her drawing. For instance, if the living room was 160 inches by 120 inches, the scale drawing would be $\frac{160}{12} = 13\frac{1}{3}$ inches by $\frac{120}{10} = 12$ inches.

Solving Single- and Multistep Problems Involving Percentages

A **percent** is a quantity per hundred. For instance, 75 percent refers to 75 out of 100. A fraction can be written in decimal form by moving the decimal point two digits to the left and dropping the percent symbol. For instance, 75 percent in decimal form is 0.75. Converting a decimal to a percentage involves moving the decimal point two digits to the right. For instance, 0.115 is equal to 11.5 percent.

Problems that involve percentages typically use the same words. For example, 45 percent of what value is 9 can be translated to an equation. The word "is" refers to the equals sign, and the word "of" refers to multiplication. The percentage is changed into decimal form and "what value" refers to an unknown quantity x. Therefore, the equivalent equation is $0.45x = 9$. Dividing by 0.45 results in $x = 20$. Therefore, 45 percent of 20 is equal to 9. Another type of problem involves finding a missing percentage. For instance, consider the problem: 11 is what percent of 110? The missing percent is the unknown quantity x. The equivalent equation is:

$$11 = x \times 110$$

Dividing by 110 results in $x = 0.1$, or 10 percent. A third problem involves finding a missing amount. For example, consider the problem: What is 8.5 percent of 90? This problem translates to the equation $x = 0.85 \times 90$, which is equal to 7.65. Therefore, 7.65 is 8.5 percent of 90. These three types of problems involve a single step to solve them.

Multistep problems can be found in real-life applications. For instance, let's say that you received a haircut for $45 and paid 10 percent sales tax on this haircut. Then, you decided to leave a 20 percent tip on the total after the sales tax. In order to find the total spent on the haircut, the amount paid with sales tax must be found first. This amount of sales tax was:

$$0.1 \times \$45 = \$4.50$$

Therefore, the total spent on the haircut before tip was:

$$\$45 + \$4.50 = \$49.50$$

The 20 percent tip was applied to this amount. Therefore, the tip was:

$$0.2 \times \$49.50 = \$9.90$$

The entire total spent on the haircut was:

$$\$49.50 + \$9.90 = \$59.40$$

Solving Problems Involving Measurement Quantities, Units, and Unit Conversion

When changing one unit to another, this process can be performed mentally or with a single step of multiplication or division if the conversion factor is known. However, the question is always whether to multiply or divide. The process of **dimensional analysis** clarifies this process. This method allows the original measurement to be multiplied by a factor of 1, so that the end result is the same quantity but in different units. Within this process, a **conversion ratio**, or **unit factor** is used. This ratio is equal to 1 and it is composed of both units needed in the conversion process.

For instance, because 12 inches is equal to 1 foot, the two unit factors associated with this relationship are $\frac{1 ft}{12 in}$ and $\frac{12 in}{1 ft}$. The one necessary to use in the conversion is the factor that has the desired end units in the numerator. For instance, if you wanted to convert 86 inches to feet, you'd use the unit factor $\frac{1 ft}{12 in}$. Multiply 86 times this factor to get the desired result.

Therefore:

$$\frac{86\ in}{1} \times \frac{1\ ft}{12\ in} = \frac{86}{12} ft = \frac{43}{6} ft$$

Notice that because there were inches in both the numerator and denominator, those units cancelled out of the expression.

Dimensional analysis can also be used in conversions that involve two steps. For instance, converting a speed in miles per hour to feet per second would involve two calculations: converting miles to feet and converting hours to seconds. However, by using dimensional analysis, this two-step process is equivalent to multiplying times two unit factors. For instance, let's say we want to convert 60 miles per hour to feet per second. We would use the fact that there are 3,600 seconds in an hour and 5,280 feet in one mile. The dimensional analysis process would be the following:

$$\frac{60\ miles}{hour} \times \frac{1\ hour}{3,600\ seconds} \times \frac{5,280\ feet}{mile} = \frac{60 \cdot 5,280\ feet}{3,600\ seconds} = \frac{88 ft}{second}$$

Given a Scatterplot, Describe how the Variables are Related

A **scatterplot** is a graph that visually represents the relationship between paired data. **Paired data** is in ordered pair form (x, y), where x is the independent variable and y is the dependent variable. Typically, the independent variable is plotted along the horizontal axis, and the dependent variable is plotted along the vertical axis, just like an ordered pair in algebra.

Here is an example of a scatterplot:

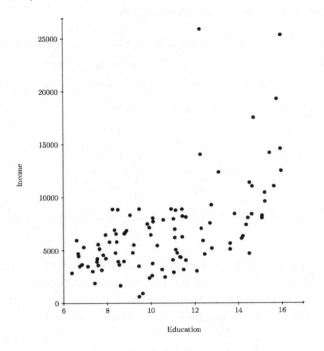

From Wikimedia Commons, by Ista Zahn,
https://commons.wikimedia.org/wiki/File:Scatterplot_of_education_and_income.png

If the data follows the shape of a straight line, a **linear model** can be used to fit the data. Other names for a linear model are a **line of best fit**, a **trend line**, or a **regression line**. It is of the form $y = mx + b$, where m is its slope, and b is its y-intercept. It is a straight line that best fits the data in a scatter plot. The line might pass through some of the points on the scatter plot, but it does not have to pass through all of them. The line of best fit is one that minimizes the distance from each point to the line.

Here is an example of a scatter plot with a linear trend line:

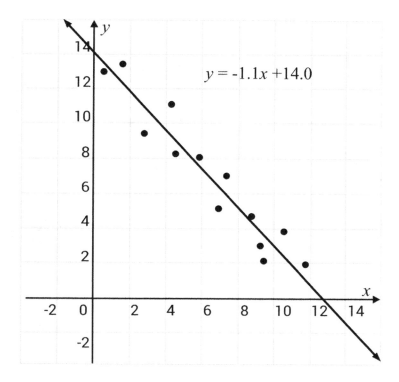

Note that the trend line passes through only two of the original data points on the scatterplot. The equation of the trend line can be used to make predictions. For instance, let's say the equation for a trend line is $y = 0.25x + 0.052$, where x represents a specific number of days of the month for a certain city, and y represents average centimeters of rain. If $x = 2$ is plugged into the equation, it results in an average of $y = 0.25(2) + 0.052 = 0.552$ centimeters of rain. In summary, a data set was used to form an equation for a trend line that can be used to estimate the total rainfall for a certain number of days in a given city.

Depending on the shape of the data in the scatterplot, other types of equations can be used to model the data. If the data in a scatterplot follows the shape of a parabola, either a "U" or an upside down "U," a **quadratic model** can be used. Technology, such as a graphing calculator or MS Excel, can be used to find the exact equation.

Here is an example of a scatterplot and a quadratic model:

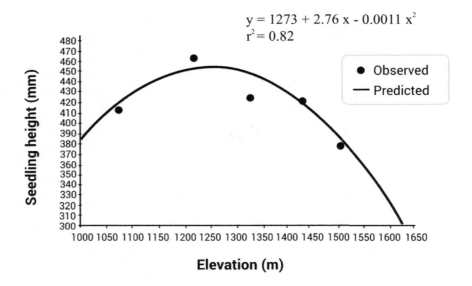

Finally, if the data in a scatterplot follows the shape of the graph of an exponential function, an **exponential model** can be used. The data can either be growing or decaying exponentially. Here is an example of an exponential model in which the data is following a pattern of exponential growth:

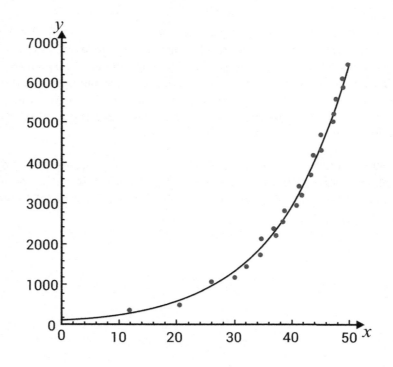

Using Two Variables to Investigate Key Features of the Graph

A scatterplot is the typical way in which paired data are graphed. Once the data are plotted, it can be seen if the data increases or decreases. For instance, if y increases as x increases, the data is increasing. If y decreases as x increases, the data is decreasing. A real-world application of increasing data would be height and weight. Typically, as a person's height increases, weight increases as well. A real-world application of decreasing data would be a person's commute length and monthly amount spent on gasoline. As a person's commute increases, the amount of money spent on gasoline would increase. Here is an example of data that increases:

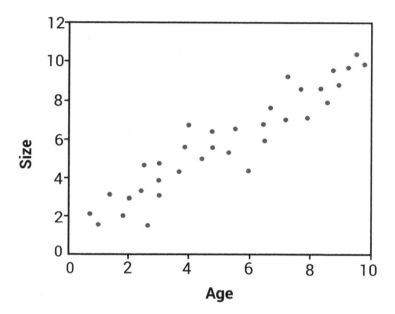

Here is an example of data that decreases:

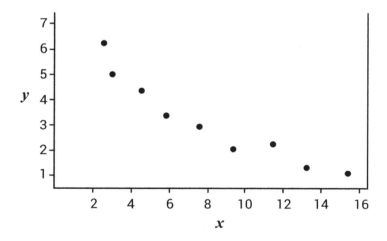

When two sets of data, as in an ordered pair, are connected, it is said that they are **correlated**. When this connection is strong, the data sets are highly correlated. Further, if the data is increasing, the correlation is positive, and if the data is decreasing, the correlation is negative. A correlation is **linear** if the data follows the shape of a line. Therefore, the first scatterplot shown above highlights a strong positive linear correlation, and the second scatterplot shown above highlights a strong negative linear correlation.

Plotted data or lines can also be used to answer questions about the data. Consider the following graph:

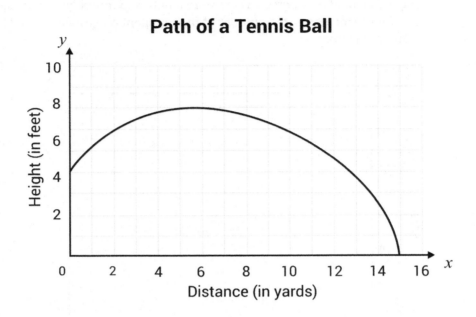

It shows the path of a tennis ball. The horizontal axis represents the distance in yards, and the vertical axis represents the height in feet. The maximum height of the ball can be found by determining the maximum value of the graph along the vertical axis. Therefore, the maximum value of the ball was 8 feet. Also, the ball was rising, or increasing, as it traveled from 0 to 6 yards away from the release point, and it was falling, or decreasing, as it traveled from 6 to 15 yards away from the release point. Finally, ordered pairs on the graph can be used to answer questions regarding the path of the ball. For instance, if we wanted to know how high the ball was when it was 10 feet away from the release point, the y —coordinate corresponding to $x = 10$ would need to be found. Therefore, 10 feet away from the release point, the tennis ball was 7 feet high.

Comparing Linear Growth with Exponential Growth

Linear growth occurs when a quantity grows the same amount given a single unit increase in input. For instance, a linear equation $y = mx + b$ represents linear growth. The rate of change, which is equal to the slope m, is constant. For every one-digit unit in x, the output y changes by a value of m. If the slope is positive, the linear growth is positive, and the graph is a straight line that increases from left to right. If the slope is negative, the linear growth is negative, and the graph is a straight line that decreases from left to right. Here is a graph that represents linear growth:

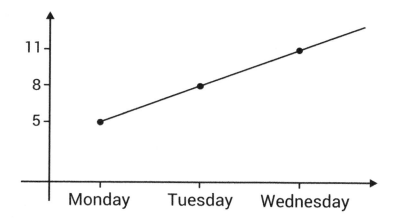

Note that for every change in day, the output variable, seen along the vertical axis, increases by 3. Therefore, the growth is constant and hence, linear.

Exponential growth is vastly different from linear growth. Linear growth has a constant rate of change. For instance, driving along a highway at a constant rate of 55 miles per hour involves linear growth. The graph of distance along a time interval would be a straight line in this scenario. The rate of change in exponential growth is far from constant. As the input variable increases, the output variable grows faster and faster. It can be said that the output variable grows "exponentially." The difference in linear versus exponential growth can also be seen in interest applications. Simple interest involves linear growth. In this setting, interest is only accrued on a starting amount. For instance, if a 1 percent interest rate for linear growth was applied to $1,000 annually, 1% of $1,000 or $10 would be added annually. Therefore, after two years, there would be $1,020 in the account. Compound interest involves exponential growth. With compound interest, the interest rate applies to both the starting amount and any accrued interest. With the same amount and rate, there would be $1,010 in the account after one year, but there would be $(1,010)(1.01) = \$1,020.10$ in the account the next time interest was calculated. As time goes on, there would be even larger monetary differences in the two types of accounts.

One equation form of exponential growth is $y = ab^x$ where a is a nonzero real number and b is a nonnegative real number greater than 1. This value b is referred to as the **growth factor**. Another way to express exponential growth in equation form is $y = a(1 + r)^x$, where r is known as the **growth rate**. Note that $1 + r = b$ and a is the value of y when $x = 0$. A real-world example of exponential growth is population growth. For instance, if a population of 25,000 people grows at a rate of 1 percent each year, the population in equation form would be $y = 25000(1 + 0.01)^x = 25000(1.01)^x$ where the independent variable x represents number of years in the future.

The graph of exponential growth is not linear. As the dependent variable increases, the independent variable grows quickly. Here is the graph of a general exponential growth equation:

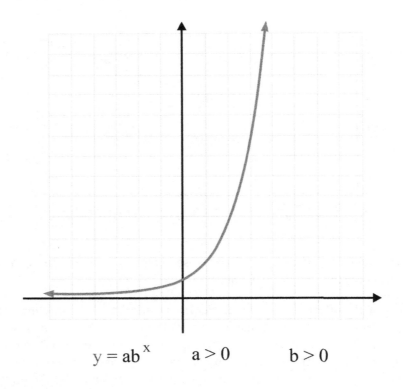

$$y = ab^x \qquad a > 0 \qquad b > 0$$

Using Two-Way Tables

A **two-way table** is a way to organize frequencies and relative frequencies for two categorical variables. A **categorical variable** is a variable that takes on a limited number of possible values. For instance, eye color would be a categorical variable because there are a limited number of possibilities. Categorical variables are nominal data because they are of a qualitative, not quantitative, nature. In the two-way table, one category is represented by the rows and the other category is represented by the columns. The table is used to highlight relationships between the two categories. The entries in the table can be either frequency counts or relative frequencies. A relative frequency of category is a fraction of the number of times that variable occurs. For example, if 4 out of 10 people have brown eyes, its corresponding relative frequency is:

$$\frac{4}{10} = \frac{2}{5} = 0.4$$

An example of a two-way table can be seen here:

	Dogs	Cats	Fish	Total
Men	5	16	8	30
Women	12	14	4	30
Total	18	30	12	60

It shows the pet preference for 30 men and 30 women, and each value outside of a total column or row represents a frequency. For example, there were 14 out of 30 women who preferred cats over dogs and fish, and 5 of 18 people who preferred dogs were men. This two-way table can be transformed into one that shows relative frequencies. Each cell value would be divided by the total number of people surveyed, which is 60. The corresponding two-way table follows:

	Dogs	Cats	Fish	Total
Men	0.0833	0.2667	0.1333	0.5000
Women	0.2000	0.2333	0.0667	0.5000
Total	0.3000	0.5000	0.2000	1.0000

The decimals are rounded to 4 decimal places. They could also be represented as percentages. Note that 0.2 or 20 percent of the people surveyed were women who preferred dogs. This amount can be thought of as a conditional probability. A **conditional probability** is a probability that an event occurs, given another event. Therefore, given that a person was a female, the probability that she preferred dogs was 20 percent. Every cell that is not in a row or column representing a total is a conditional probability.

Make Inferences about Population Parameters Based on Sample Data

There are two types of statistics: descriptive and inferential statistics. **Descriptive statistics** is used to summarize the basic features of a data set. This type of statistic involves calculations. For example, finding the median of a data set is part of descriptive statistics. **Inferential statistics** is the other branch of statistics, and it involves making inferences about a population using sample data. In this context, a population contains all subjects being studied. For instance, a population could be all male adults over the age of 18 in a certain country. Data from this entire set of people, like each person's individual weight, would be extremely hard to obtain. Because of this difficulty, sample data would be used to make decisions about this population. A **sample** consists of a smaller subset from a population. A corresponding sample would be fifty males from this country. The weights of fifty people would be much easier to obtain. Specifically, inferential statistics involves the use of a sample to generalize about populations. Therefore, one could find the mean weight of those fifty individuals to generalize the mean of the entire population.

Hypothesis testing is the formal process for assessing claims made about a population based on a sample. A **hypothesis** is the actual claim regarding a calculation from a population. A calculation from a population, such as the mean weight of every person in that group, is difficult to obtain and is known as a **parameter**. A hypothesis may or may not be true, and sample data is used to make that decision. The **null hypothesis**, H_0, and the **alternative hypothesis**, H_1, are the two types of hypotheses used in a hypothesis test. The null hypothesis states that there is no difference between two parameters, and the alternative hypothesis states that there is a difference between two parameters. For our weight example, if the male population in another country had a mean weight of 198.5 pounds, a researcher might want to test if the population's mean weight in her country was less than that amount. She calculates the mean of the fifty males in her sample data and conducts a hypothesis test. The null hypothesis states that the mean weight of males in her country is equal to 198.5 pounds, and the alternative hypothesis states that the mean weight of males in her country is larger than 198.5 pounds. A statistical test would then involve using the sample data to determine whether the null hypothesis should or should not be rejected.

Measures of Center of Data and Shape, Center, and Spread

In statistics, a measure of center is a measure of average of a data set. They are also known as **measures of central tendency**, and include the mean, median, mode, and midrange calculations. The **mean**, or arithmetic average, is calculated by dividing the total of the data by the total number of data points in the data set. For instance, the mean salary of 100 people is found by adding all 100 salaries and dividing by 100. The **median** is the midpoint of all the data in the set. If there is an odd number of data points, the median is equivalent to the data point exactly in the middle. If there is an even number of data points, there are two data points in the middle of the data, and the median is equal to the mean of these two values. Therefore, the median of the 100 salaries is equal to the mean of the two salaries in the middle of the data when written in numerical order. The **mode** is equal to the value that appears the most in a data set. In other words, it has the highest frequency. Therefore, the mode of the 100 salaries would be the salary that appeared the most times. A data set with one mode is referred to as **unimodal**. A data set can have no mode. In this case, each value would only appear one time. It can also be **bimodal**, having two modes, or **multimodal**, having more than two modes. Finally, the **midrange** is the mean of the lowest and highest data points. In the salary example, the midrange is equal to the smallest salary plus the largest salary divided by 2.

The best measure of center in a data set is distinct for different types of data sets. If a data set has many outliers, data values that fall far away from most of the other data points, the mean calculation would be heavily affected. For instance, if the salary example consisted of 99 normal employees and the CEO, where the CEO's pay is much higher than the others, this large salary would skew the mean to be quite higher because the sum of all salaries would include this large amount. In this case, the median would be a better measure of center for this data set because it is unaffected by this outlier. The midrange is also skewed by outliers because it consists of a direct calculation with the highest value in the data set. For categorial data, the mode is always the best measure of center because calculations cannot be made with this type of data.

A **data set** can be described using its standard deviation, or its spread. This calculation quantifies how spread apart the data is, meaning how much the data vary with respect to its mean. A standard deviation closer to 0 shows that most of the data in the data set are close to the mean. A standard deviation equal to 0 shows that all values in the data set are equal. A standard deviation further from 0 shows that the data set varies largely from the mean. Here are two distributions that show data with both a smaller standard deviation and a large standard deviation:

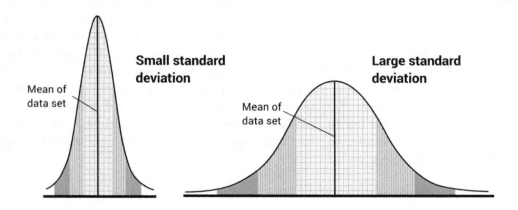

In the graphics, the solid black line represents the mean of the data set. Notice in the small standard deviation example, the majority of the data are close to this line, and in the large standard deviation example, the majority of the data are not as close.

Evaluating Reports

For inferential statistics to be reliable, a sample must be obtained in a smart manner. For instance, if a researcher wanted to find out the percentage of a population that prefers tea over coffee, they would not want to obtain a sample by surveying people at a tea house. This decision would probably result in a **biased sample**, meaning that one preference would be more likely than the other. A sampling done at a teahouse would probably result in a higher percentage of people preferring tea over coffee. A good sample is unbiased. In the coffee example, the researcher might want to head to a grocery store and select every tenth person who walks in or send out an online survey that asks a question about preferences. These options are called **sampling methods**. Sampling methods are not perfect, and each type has its own advantages and disadvantages.

There are five types of sampling methods: random, systematic, convenience, cluster, and stratified. **Random sampling** occurs when every member of the population has an equal chance of being selected for the sample. Computer generated lists can be used within random sampling. A random number generator can be used to give a random number to each member of the population. In this method, a sample is obtained by selecting a finite number of random numbers, say 30, and utilizing only those members of the population in the sample.

A **systematic sample** is when a sample is obtained by counting off members of the population. Every kth member of the population is used. For instance, the grocery store example listed previously is a systematic sample.

A **convenience sample** is found when only data that is readily available, or convenient, is used. In most cases, this type of sampling method should not be used. An example of this method would be the tea house example. It tends to lead to biased samples.

A **cluster sample** is obtained by using an already designed sample. For instance, schools are divided into classrooms. Instead of using the entire fourth grade, one classroom could be selected as a sample. Another example would be selecting an entire zip code to represent a sample of a metropolitan area.

A **stratified sample** is obtained when the population is broken up into **strata**, or subsets, based on a characteristic like age, and then, members from each stratum are selected randomly. An example of this would be a city broken up into age ranges and a survey sent to 100 people in each age group.

Once a sample is obtained, data need to be collected and processed. A survey, as previously discussed, is a common way to obtain data. Other methods involve interviews, observations, and focus groups. Not all data collection methods work for every type of sample. For instance, if a sample consisted of people spread across the globe, focus groups probably would not be convenient. In this case, surveys might work better because they would be quicker and less expensive to implement.

Passport to Advanced Math

Creating a Quadratic or Exponential Function or Equation that Models a Context

A **quadratic equation** is a model of a parabola; **parabolas** are U-shaped curves that are said to be concave up (like a regular "U") or concave down (an upside-down "U") depending on the sign of the coefficient on squared term in the quadratic. When the leading term has a positive sign, the parabola is concave up, resembling a standard letter "U." A common real-world example of a parabola is the path of a projectile. When golf balls are hit or baseballs are thrown, the path that they take is a parabola, which can be modeled by a quadratic equation. Quadratic equations are written in three different forms: standard, intercept, and vertex form. Each of these forms gives different information about the graph of the parabola. For vertex form, $y = a(x - h)^2 + k$, the vertex is found at (h, k) and the axis of symmetry is $x = h$. The value of a tells you whether the graph is concave up, in which case a is positive, or concave down, in which case a is negative. For the graph of balls thrown into the air, the y-value typically represents the height, and x-value can be the time since the ball was thrown. The value of a is negative, and it represents the pull of gravity. The vertex is the highest point that the ball reaches before beginning to fall to the ground.

Another form for quadratic equations is intercept form:

$$y = a(x - p)(x - q)$$

The information that this form gives is the x- and y-intercepts. The y-intercept is the initial height from which a projectile is thrown. The x-intercept is the time it takes for the projectile to reach the ground.

The most common form of a quadratic equation is standard form:

$$y = ax^2 + bx + c$$

The value of a represents the concavity of the graph. To find the axis of symmetry, you use the equation $y = \frac{-b}{2a}$. Using this outputted y-value from the axis of symmetry equation, the vertex is found by solving for the corresponding x-value.

One example of a context that can be modeled by a parabola is the path of a shot-put throw. Using the following equation, there are many characteristics that can be seen:

$$y = -1.5x^2 + 7x + 1.5$$

The y-value represents the height of the shot, while the x-value is the time that passes during the throw. The coefficient of x tells you the initial velocity of the shot. The c-value, or 1.5, is the initial height that the shot was pushed from. The following equation will find the vertex:

$$y = \frac{-b}{2a} = \frac{-7}{2(-1.5)} = 2\frac{1}{3}$$

This value ($2\frac{1}{3}$) represents the maximum height of the shot during the throw. The time it took for the shot to reach this point is found by evaluating the equation $\frac{7}{3} = -1.5x^2 + 7x + 1.5$ to solve for x.

Quadratic equations can be written to model real-life contexts. Suppose that a ball is thrown into the air at an initial height of 4 feet, with an initial velocity of 10 meters per second. You can write a quadratic equation to model the path of the ball. With the given information, the standard form of the equation is written as:

$$y = ax^2 + 10x + 4$$

The missing value of a is found to be $-\frac{1}{2}(9.81)$, where 9.81 is the pull of gravity in units of meters per second squared. With this information, the equation is solved as:

$$y = -4.9x^2 + 10x + 4$$

Another situation that can be modeled by a quadratic equation is the path of a dolphin's jump over the water. The equation $y = -0.3x^2 + 4x$ models the jump in terms of x, the time of the jump, and y, the height of the jump. To find the point in time when the jump reached its maximum height, use the formula:

$$x = \frac{-4}{2(0.3)} = 6\frac{2}{3}$$

To find the jump's maximum height, evaluate the equation for y at this calculated value of x.

While the path of a projectile are often modeled by a quadratic function, other real-life scenarios are modeled by exponential functions. Exponential functions represent either exponential growth or exponential decay. Unlike quadratic functions, exponential functions do not have a vertex. They either increase or decrease at a changing rate. Exponential functions take the form of $y = ab^x$. If the value of b is greater than one, it represents exponential growth. If the value of b is between zero and one, the function represents exponential decay.

The following table represents values that are consistent with an exponential function.

x	-2	-1	0	1	2	3
y	$\frac{1}{4}$	$\frac{1}{2}$	1	2	4	8

The y-values increase as the x-values increase. Since the outputs are increasing, the table represents exponential growth. The change in y-value from one input to the next is different. Because each output is a factor of 2, this function is exponential and takes the form of the equation $y = 2^x$. To find the next value, evaluate the function at $x = 4$. The output is $y =$ 16. The exponential growth is seen in how the changes in y increase by a factor of 2 each time.

Another example that can be modeled by an exponential function is the growth of bacteria. The following table shows bacteria growth over a period of five days.

x	0	1	2	3	4	5
y	3	12	22	50	82	113

From the changes in the y-values, this function can be seen as exponential growth. Using a graphing calculator, these values can be input into the lists and an equation modeled. The following equation models the table:

$$y = 4.6 \times 2.0^x$$

The value of a is 4.6, which represents the y-intercept. The y-intercept from the table is 3, but since this is a model of the situation, there will not be an exact match to the data. The base value of 2 is the approximate factor by which the y-values increase. This equation can be used to estimate the growth of the bacteria in the days following day 5.

Suitable Forms of Expression

In the following equation, $y - 2 = 2(x - 3)$, two traits of the function can be seen, the slope and a point. The slope of this line is 2, and a point on the line is $(2, 3)$. To find other characteristics like the y-intercept, you manipulate the equation. Solving for y yields the equation $y = 2x - 4$. In this slope-intercept form of the equation, the y-intercept is found to be -4. These same concepts can be used to find characteristics of other equations.

The equation $y = -x^2 + 2x + 4$ is written in standard form. This form shows that the parabola is concave down because of the coefficient of x^2, and the y-intercept is 4. This form does not reveal the x-intercepts. Factoring the equation puts it in intercept form as:

$$y = -(x - 2)(x + 2)$$

From this form, the x-intercepts are seen to be $x = 2$ and -2.

In a similar way, the following equation gives some characteristics of a situation:

$$y = -(x - 3)(x + 3)$$

If the value of y represents the height of a balloon, then x represents the time in minutes after the balloon left the ground. From the intercepts, it can be seen that the balloon landed after 3 minutes in the air. To find the y-intercept, or the height at which the balloon was released, change the equation. Simplified, the equation becomes:

$$y = -x^2 + 9$$

From this form, it is apparent that the initial height of the balloon is 9 feet.

Equivalent Expressions Involving Rational Exponents and Radicals

Equivalent expressions are those that are written in different forms but are equal to each other. Rational exponents are those exponents that can be written in the form of fractions. Examples of rational exponents are $n^{\frac{3}{4}}$ and $a^{\frac{1}{2}}$. These expressions can be written using radicals instead of fractional exponents. For a fraction in the exponent, the denominator becomes the root and the numerator is the exponent. The expression $n^{\frac{3}{4}}$ becomes the fourth root of n to the third power, $\sqrt[4]{n^3}$. An equivalent expression can also be written for $a^{\frac{1}{2}}$. The root of a is 2 and the exponent is 1. This expression written with a radical is $\sqrt[2]{a^1}$, or $\sqrt{a}$. Any time the denominator on the exponent's fraction is 2, the root is a square root.

An expression can also be written in a radical form first. The expression $\sqrt[5]{x}\sqrt[5]{y}$ can be rewritten as $x^{\frac{1}{5}}y^{\frac{1}{5}}$. Once the expression is transformed by using rational exponents, it can be combined to form $(xy)^{\frac{1}{5}}$. It is helpful to be able to transform these equations so that equivalent expressions can be used to evaluate different situations.

The following problem involves multiple variables with different exponents: $\frac{x^3 x^{-4}}{xy^{-4}}$. These variables can be simplified one at a time. The variable x can be combined in the numerator to form the term x^{-1}. The new expression is $\frac{x^{-1}}{xy^{-4}}$. When you take the variables that are raised to a negative power and then find the reciprocal, the expression becomes $\frac{y^4}{xx^1}$. Combining both x terms makes the simplified expression $\frac{y^4}{x^2}$.

For some expressions, there are both radicals and rational exponents expressed in different forms. Take the following expression, for example: $\frac{\sqrt{x}}{yx^3}$. The numerator can be written as $x^{\frac{1}{2}}$. You use the rule of subtracting exponents when the bases are the same, and then you divide the numerator and denominator; the x-term becomes $x^{-\frac{5}{2}}$. The exponent is negative because the exponents subtracted were:

$$\frac{1}{2} - 3 = -2\frac{1}{2} = -\frac{5}{2}$$

When you simplify the expression, the result is $\frac{1}{y\sqrt{x^5}}$, because $x^{\frac{5}{2}}$ is the same as the square root of x to the fifth power, $\sqrt{x^5}$.

Equivalent Form of an Algebraic Expression

An **equivalent form of an algebraic expression** is one that is equal to the original expression but written in a different way. Depending on what the expression represents, equivalent forms may be helpful if they give information based on different variables. For example, the expression $15x + 25$ may represent the total price that 5 friends pay to visit the movies. If you want to know the price per friend, then the factored form $5(3x + 5)$ would show the price for each of the friends to be $3x + 5$. These equivalent forms can be found by manipulating expressions. One form of the expression is $2(4x + 5y)$, and an equivalent form is $8x + 10y$. To find this latter form, you distribute the 2 to each term inside the parenthesis.

Another way to manipulate expressions is by factoring out a number. Suppose that the original expression is $\frac{4xy}{2y}$. You can find an equivalent form by factoring out the common values of $2y$ from the numerator and denominator. This equivalent expression is $(2y)\frac{2x}{1}$, or $2y(2x)$.

Expressions can be rewritten in different ways. Take the following expression, for example:

$$\frac{t^2 - 7t + 12}{t - 4}$$

The numerator can be factored into:

$$\frac{(t-4)(t-3)}{t-4}$$

When you find the equivalent form of this expression, there are two binomials on the top and bottom that can help simplify the equation.

Solving a Quadratic Equation

Quadratic equations can be solved by using different methods. When the quadratic equation is simple and has no single x terms, the equation can be solved by taking the square root. For example, the equation $4x^2 = 16$ can be solved by using the rules of algebra. First, divide the 4 on both sides to give the equivalent equation $x^2 = 4$. Use the square root to find the values of x that make the equation true. The solution to this equation is $x = 2, -2$.

Another method of solving quadratic equations is by factoring. This method works for equations in standard form with certain coefficients of x^2. Take the following equation as an example:

$$x^2 - 5x + 6 = 0$$

The factored form of this equation is:

$$(x-2)(x-3) = 0$$

This form enables the values of x to be found as $x = 2, 3$.

Quadratic equations can also be solved by using the quadratic formula. For this method, the equation must start in standard form:

$$y = ax^2 + bx + c$$

The quadratic formula is:

$$x = \frac{-b \pm \sqrt{b^2 - 4ac}}{2a}$$

The values of a, b, and c from the standard form of the equation can be plugged in to the formula, and the equation can be solved for x.

Another method for solving quadratic equations is by **completing the square**. The equation must start in standard form:

$$ax^2 + bx + c = 0$$

The first step is to isolate the constant value by moving it to the right side to form:

$$ax^2 + bx = -c$$

Now, complete the square by finding half of the constant b, squaring it, and adding this value to both sides. Next, you factor in the left side:

$$(x + \frac{b}{2})^2$$

For example, take the following equation:

$$x^2 + 8x - 4 = 0$$

Isolate the constant, which results in the equation:

$$x^2 + 8x = 4$$

Now complete the square: take half of 8 (which is 4), square it (which is 16), and add it to both sides of the equation. The new equation is:

$$x^2 + 8x + 16 = 20$$

This new equation can be factored to read:

$$(x + 4)(x + 4) = 20$$

Written using exponents, it becomes:

$$(x + 4)^2 = 20$$

The next step is to solve for x by isolating the variable. Taking the square root of both sides gives the equation:

$$(x + 4) = \sqrt{20}$$

Subtracting 4 from both sides gives the two solutions:

$$x = -4 \pm \sqrt{20}.$$

Adding, Subtracting, and Multiplying Polynomial Expressions

A **polynomial expression** is an expression in terms of a variable, x, with non-negative powers that are integers. These expressions can be simplified when they have two terms that are considered "like terms." For example, the terms $2x^2$ and $4x^2$ are like terms because they have the same variable, x, with the same degree, 2. These terms can be added by finding the sum of the coefficients to yield a final expression:

$$2x^2 + 4x^2 = 6x^2$$

Other examples of adding polynomial expressions can be seen in the following terms:

$$2x^2 + 8x + 3 - 5x^2 - 3x$$

In this expression, there are two sets of like terms that can be collected into one final expression:

$$-3x^2 + 5x + 3$$

For more complex polynomials, the same rules apply for adding and subtracting like terms. The following expressions can be added:

$$(8 + 4x^2 - 4x) + (-3x^2 + 12)$$

By collecting like terms, the two expressions can be combined into one: $x^2 - 4x + 20$. These final terms cannot be combined because the degrees of the variable are different.

Two polynomial expressions can also be subtracted. The following expressions can be collected after distributing the negative:

$$(8r - 5r^2 - 10) - (-3r^2 + 5r)$$

The expressions become the addition of $(8r - 5r^2 - 10) + (3r^2 - 5r)$ where the final result is:

$$-2r^2 + 3r - 10$$

The key to subtracting polynomials is to distribute the negative through the second expression.

Another way of manipulating expressions is to multiply them. The following expressions can be multiplied as $(x - 3)(x + 2)$. When two binomials are multiplied, use the rule of FOIL (first, outer, inner, last). The first terms are multiplied, then the outside terms, then the inside, then the last. After multiplication, the expression becomes:

$$x^2 + 2x - 3x - 6$$

Then collect like terms to have a final expression of $x^2 - x - 6$. Other polynomials can be multiplied as well. Take the following expressions, for example:

$$(t - 3)(t^3 - t + 2)$$

To multiply these expressions, each term in the first polynomial must be multiplied by each term in the second. The resulting polynomial is:

$$t^4 - 3t^3 - t^2 + 2t - 6$$

Solving an Equation in One Variable that Contains Radicals

Solving an equation in one variable means that variable needs to be isolated. You find the value of the variable by isolating the variable. Depending on the complexity of the equation, this process of isolating the variable can be done in one step or many steps. For example, take the following equation involving a radical: $\sqrt{x} = 4$. In order to isolate this variable, you must remove, or cancel, the radical above the x. To remove the radical, you must use the opposite operation. The opposite operation of a square root is a square. By raising both sides of this equation to the second power, you remove the radical and isolate the variable. The answer is $x = 16$.

In a similar equation involving radicals, the following x-value requires more operations to be isolated: $\frac{\sqrt{x^4}}{4} = 1$. The first step is to move the 4 to the right side by opposite operations. Then multiply both sides by 4; the equation becomes $\sqrt{x^4} = 4$. Now you must remove the radical by performing the opposite operation of a square root. By raising both sides to the second power, you make the equation $x^4 = 16$. Now you must remove the power of the x to find the value of the single variable. The opposite of raising a variable to the fourth power is taking the fourth root, or raising both sides to the $\frac{1}{4}$- power. When the variable is isolated, the x-value is found to be $x = 2$.

The previous two examples involved only a monomial under the radical. The steps to solving an equation may be more complex when there is a polynomial under the radical. For example, take the following equation:

$$3\sqrt[4]{x + 2} = 6$$

The first step in isolating the variable is to isolate the radical. Divide both sides by 3, so that the equation becomes:

$$\sqrt[4]{x + 2} = 2$$

Next, take the opposite of a fourth root by raising both sides to the fourth power. The resulting equation is $x + 2 = 16$. Subtract 2 from both sides of the equation; the value of x is 14.

Another type of single-variable equation is an equation in which the variable is found in the denominator. For example, take the following equation: $\frac{10}{x} = 5$. Because the x is on the bottom of the equation, the equation can also be written as:

$$10x^{-1} = 5$$

In order to cancel out the 10, divide both sides of the equation by 10 to yield the equation $x^{-1} = \frac{1}{2}$. Raise both sides of the equation to -1 to isolate the x-value. This yields the answer of $x = 2$.

Another example of the variable in the denominator is the following equation:

$$\frac{5}{x - 4} = 15$$

Again, the first step in this equation is to rewrite it using exponents:

$$5(x - 4)^{-1} = 15$$

Divide both sides of the equation by 5, which isolates the variable:

$$(x - 4)^{-1} = 3$$

Raising both sides of the equation to the power of -1 yields the equation:

$$x - 4 = \frac{1}{4}$$

The final step is to add 4 to both sides, yielding the answer $x = 4\frac{1}{4}$.

Solving a System of One Linear Equation and One Quadratic Equation

To find a solution to a system of equations, you find the place where they intersect. Some systems of equations have more than one solution. For a system of one linear equation and one quadratic equation, there could be zero, one, or two solutions.

For example, take the following equations: $y = 2x - 1$ and $y = x^2$. Use the substitution method to set these two equations equal to each other, forming one equation:

$$x^2 = 2x - 1$$

After you work to get all the terms on one side, the equation becomes:

$$x^2 - 2x + 1 = 0$$

You can then factor the quadratic equation into two binomials:

$$(x - 1)(x - 1) = 0$$

Rewrite the equation as $(x - 1)^2 = 0$. Next isolate the value of the variable by taking the square root of both sides. The final step in solving the equation is adding 1 to both sides. The value of the variable is 1. This number is the x-value part of the solution. To find the y-value of the solution, substitute $x=1$ into either equation. The first equation, $y = 2x - 1$, becomes $y = 2(1) - 1$, where the y-value equals 1. The solution, or the place where the lines intersect, is $(1, 1)$.

Another example of a system of equations with one linear and one quadratic equation is $y = 4x - 2$ and $y = x^2 + 2$. Again, set both of these equations equal to y, so they can be substituted, which forms the following equation with one variable:

$$4x - 2 = x^2 + 2$$

Using opposite operations, you can solve the equation for zero:

$$x^2 - 4x + 4 = 0$$

After you factor this quadratic into two binomials, it becomes:

$$(x - 2)(x - 2) = 0$$

When this equation is simplified, write it as:

$$(x - 2)^2 = 0$$

Take the square root of both sides and solve for x; this part of the solution is $x = 2$. With this information, substitute the x-value into either original equation to find the y-value. The equation $y = 4x - 2$ can be used to yield:

$$y = 4(2) - 2 = 6$$

The final solution, or the one point where the two lines intersect, is $(2,6)$.

Rewriting Simple Rational Expressions

Rational expressions can be rewritten, depending on which variables are involved and what their coefficients have in common. Take the following expression, for example:

$$\frac{5x + 10}{40x + 15}$$

The binomial in the numerator has a common factor of 5 in each term. Factoring that out to the front yields the expression $5(x + 2)$. The denominator also has a common factor of 5 in the two terms. Factoring this factor out to the front yields the expression $5(8x + 3)$. Writing the expression again yields the following expression with the factors out front:

$$\frac{5(x + 2)}{5(8x + 3)}$$

Since $\frac{5}{5}$ cancels to 1, the original expression can be written as:

$$\frac{(x + 2)}{(8x + 3)}$$

Multiplying two rational expressions sometimes requires the use of the rules of exponents. For example, the expression $\frac{3x^2}{9} \times \frac{3}{x}$ can be simplified by multiplication first. Multiply the numerator to form $9x^2$. The denominator becomes $9x$. Since there is a common variable on the top and bottom, subtract the exponents to form x. The coefficients of x are the same number, 9, so they cancel to give the final expression x as the simplified form.

Rational expressions can also be added and subtracted. The rules for adding fractions are like those required to add and subtract rational expressions. First, the least common denominator (LCD) must be found. Take the following expressions, for example:

$$\frac{7 + x}{(x - 2)} + \frac{4}{(x + 1)}$$

Since the denominators are not the same, the LCD is the product of $(x - 2)$ and $(x + 1)$. Taking what is missing from each fraction and multiplying it by the numerator and denominator yields the expression:

$$\frac{(7 + x)(x + 1)}{x^2 - x - 2} + \frac{4(x - 2)}{x^2 - x - 2}$$

If the numerators are simplified, they become $x^2 + 8x + 7$ and $4x - 8$. Combining like terms, since the denominators are the same, gives the final expression:

$$\frac{x^2 + 12x - 1}{x^2 - x - 2}$$

Similar rules apply for subtracting rational expressions. Take the following expression, for example: $\frac{3}{x} - 4$. In order to rewrite this expression, find the LCD, which is x. Multiply this variable by the numerator and the denominator, which makes the second term $\frac{4x}{x}$. Now that the expressions have common denominators, combine the numerators to form the expression:

$$\frac{3 - 4x}{x}$$

Interpreting Parts of Nonlinear Expressions in Terms of their Context

The difference between linear and nonlinear equations is in the relationship between the independent variable and the dependent variable. In a linear equation, the output value changes at a constant rate in relation to the input. In a nonlinear equation, the output related to the input is not constant. Examples of linear relationships are equations such as $y = 2x - 4$, $y = -4x$, and $y = x + 3$. Nonlinear relationships may be represented by quadratic equations such as $y = 2x^2$ or exponential equations such as $y = 4^x$. Other examples of nonlinear relationships are inverse equations, radical equations, and absolute value equations.

Interpreting parts of nonlinear equations requires knowledge of the relationship between the commonly used variables x and y. A quadratic equation is often written in the form:

$$y = ax^2 + bx + c$$

When this equation is used to model the position of an object in relation to time, the following things can be interpreted from the equation. Take the quadratic equation:

$$y = -16x^2 - 30x + 200$$

The coefficient of x^2, which is 16, tells you that the pull of gravity is 16 feet per second to the ground. The coefficient of x is -30, which tells you the initial velocity of the object. The constant number at the end of the equation tells you the initial height that the object is thrown from. Each part of the equation shows different information that models the object. The equation is nonlinear because the path of an object effected by gravity is not a straight line. Since the degree, or exponent, of the x is 2, the relationship is a quadratic.

Another type of nonlinear expression is an exponential one. An exponential equation is one in which the input value, normally x, is found in the exponent position. The standard form of an exponential equation is $y = ab^x$. Take the following equation as an example: $y = 3(2^x)$. The value of 3 shows the y-intercept value if the equation were graphed at (0,3). The value of 2, or the base that is raised to an exponential power of x, shows that the equation represents exponential growth. In contrast, if the value of b were less than 1, then the equation would represent exponential decay.

Relationship Between Zeros and Factors of Polynomials

You can factor polynomial equations if their terms can be broken down into two or more parts. For example, the following equation is a polynomial:

$$y = 2x^2 - 4x$$

Finding the factors of the polynomial involves looking for what the two terms have in common. For this equation, the terms have $2x$ in common. Factoring this out to the front gives the equation:

$$y = 2x(x - 2)$$

To find the zeros of this equation means finding the values of x that make the y-value equal to zero. When the equation is set equal to zero, it is:

$$0 = 2x(x - 2)$$

By the multiplicative property of zero, if one factor is equal to zero, the product is zero. By applying this property to the given equation, the solutions can be found. Take these two equations as the next step, $2x = 0$ and $x - 2 = 0$. Solving each equation for x yields the two values $x = 2$ and $x = 0$. These two values are the solutions to the equation. In other words, these two values are where the line crosses the x-axis.

Another example of the relationship between zeros and factors can be found in the following equation:

$$y = x^2 - 6x + 5$$

These terms do not have anything in common to factor out to the front. The next step in looking for factors is finding two binomials whose product is the polynomial given. This trinomial can be factored into:

$$y = (x - 5)(x - 1)$$

Set this equation to zero, which gives the solutions $x = 5$ and $x = 1$. The factors from the equation yield the values that make $y = 0$. These are the solutions to the equation, or the intersection of the line and the x-axis.

Another example of a polynomial equation with factors that yield multiple solutions is shown in the following graph. The equation $y = x^2 - x - 12$ is a parabola. This line crosses the x-axis at two points, $x = -3$ and $x = 4$. To see the connection between factors of the polynomial and these solutions, you must break down the polynomial. When you rewrite the equation with two binomials, it becomes:

$$y = (x + 3)(x - 4)$$

Set the y-value equal to zero, and then rewrite the equation as:

$$0 = (x + 3)(x - 4)$$

When you solve this equation for x, the values turn out to be the same as those points where the parabola crosses the x-axis.

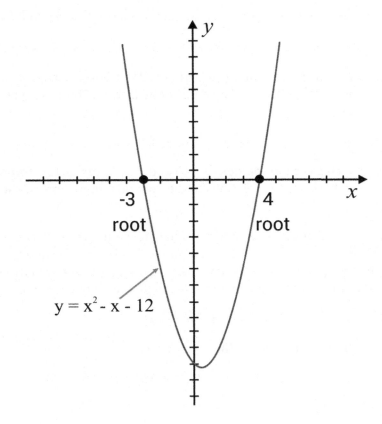

For equations with degrees higher than 2, the number of solutions is equal to the degree. For example, the equation $y = x^3 - 4$ has a degree of 3, and this equation has three solutions. The same is true for higher degrees. The equation $y = 5x^5 - 3x^4 + 4x^3 - 9x^2 + 2x - 3$ has a degree of 5, and it will have 5 solutions.

Nonlinear Relationship Between Two Variables

A nonlinear relationship between two variables is more complicated than a linear relationship, in which the variables are proportional. For example, the equation $y = x^2$ represents a relationship that forms a parabola when graphed. As the x-value, or input, increases, the y-value increases by the square of the input. Another type of nonlinear equation is a cubic equation. An example of this type of equation is shown in the graph below. When the x-value is less than zero, the y-value is less than zero. As the x-

value decreases by one, the y-value decreases by three times that number. The same is true on the right side of the y-axis. As the x-value increases by one, the y-value increases by three times that amount.

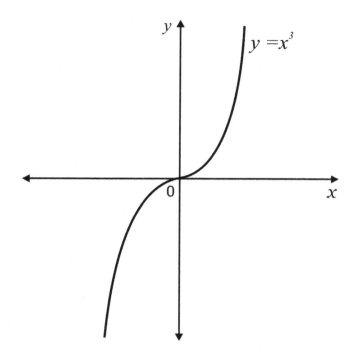

Another nonlinear relationship between two variables can be seen in the equation $y = \frac{1}{x}$. When the x-value is 1, the y-value is 1. As the value of x increases, the value of y increases by a fraction of that input amount. When x is less than 1, but greater than zero, the y-value increases by the reciprocal of that fractional amount. This relationship can be seen in the modeling of speed, distance, and time. The common equation relating these three variables is $speed = \frac{distance}{time}$. This type of relationship can show how time changes in relation to speed for a given distance. If the speed increases, then the time decreases for a constant distance traveled.

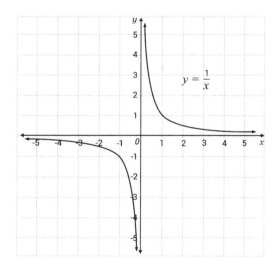

Function Notation

Function notation is a way to name a function that is defined by an equation. Most functions have the variable x as the input value and f of that variable (such as as $f(x)$) as the name of the function. An example of a function is $f(x) = 3x - 4$. In this notation, the equation can be read as "f of x is equal to $3x - 4$". Using function notation can be a way to evaluate a function for a given value of x. With the function $f(x) = -2x + 5$, the value at $x = 3$ can be found by the equation:

$$f(3) = -2(3) + 5$$

Evaluating this function yields the value $f(3) = -1$.

The defining characteristic of every function is that for every input value, there is one and only one output value. Function notation cannot be used if this rule is not followed. For example, the following values in the table on the left represent a function denoted by an equation written in function notation. The values presented in the table on the right do not represent a function because there are two different output values for the input value of 4.

Function		Not a function	
Input	Output	Input	Output
-1	5	3	0
0	3	4	7
1	4	5	10
2	7	4	14
3	4	10	25

Another example of function notation used in context is an equation that shows the distance that a migrating bird travels in a given number of days. Let us suppose the equation is $f(t) = 50t$. To use this equation in context is to evaluate how far the bird migrates each day. After five days, the total distance traveled can be found by the equation:

$$f(5) = 50(5)$$

When this equation is evaluated, it becomes clear that the distance that the bird traveled is 250 miles. This same method can be used to calculate the total distance traveled after any number of days.

Using Structure to Isolate or Identify a Quantity of Interest

The following image shows the equations that are used to find a circle's area, A, and its circumference, C. In some situations, these equations can be used in the given form to find the unknown value. In other situations, given different known values, the equation must be manipulated to find the unknown value. For example, the first equation for area can be changed in order to find the value of the radius, r. The first step is to divide by π, which makes the equation $\frac{A}{\pi} = r^2$. The last step is to isolate the r-value by taking the square root of both sides, making the equation $\sqrt{\frac{A}{\pi}} = r$. Each step in this process is made possible by using the properties of equality. When one operation is done to one side, the same operation must be done to the other side.

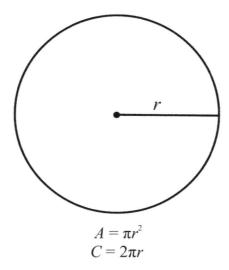

$A = \pi r^2$
$C = 2\pi r$

Another example of using structure to identify a quantity of interest is the following equation for volume. Imagine a problem situation that provides values for volume, length, and width, but the height is missing. In order to solve for the variable h, use the properties of equality on the equation. The first step in isolating h is to multiply by 3 in order to cancel out the $1/3$.

Then the equation becomes $3V = lwh$. The next steps can be combined by dividing by lw, which makes the equation $\frac{3V}{lw} = h$. This equation now yields itself to the given knowns of volume, length, and width, while solving for the unknown h.

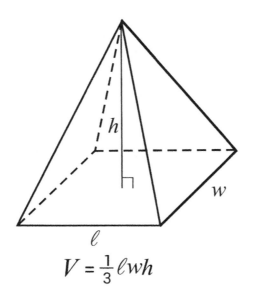

$V = \frac{1}{3}\ell w h$

Practice Quiz

1. The table below gives the initial population of a species (at time $t = 0$), and the population at each year for 3 years.

Time (years)	Population
0	2,000
1	8,000
2	32,000
3	128,000

Which of the following functions best models the population $P(t)$ at year t?
   a. $P(t) = 4,000t + 2,000$
   b. $P(t) = 4,000t$
   c. $P(t) = 2,000(4^t)$
   d. $P(t) = 2,000(4^{-t})$

2. A housing developer builds twelve new houses a year. Which of the following types of functions best models the number of houses being built as a function of time t.
   a. Exponential growth
   b. Exponential decay
   c. Increasing linear
   d. Decreasing linear

3. A pizza delivery driver worked last Saturday night. The amount of money that he made, M, after hour h can be modeled by the equation $M = 7.45h + 16$. What is the best interpretation of the number 7.45 in the context of this problem?
   a. The number of pizzas delivered
   b. The number of hours worked
   c. The total amount of money made on Saturday
   d. The hourly wage of the driver

4. The mean of 7, 19, x and 54 is 30. What is the value of x?

5. Chris works for an internet provider. Each week, he receives a list of houses in which he needs to complete an installation. The number of houses that he needs to visit at the end of each day can be modeled by the equation $H = 208 - 26d$, where H is the number of houses left for installation and d is the number of days that he has worked that week. What is the meaning of the value 208 in this equation?
    a. It represents the number of houses Chris visits each day.
    b. It represents the number of houses Chris must visit each week.
    c. Chris needs 208 days to complete all of his installations.
    d. Chris charges $208 per installation.

Answer Explanations

1. C: The options listed are both linear and exponential. Because population is not growing at a constant rate, the model must be exponential. The population is growing, and $P(t) = 2,000(4^t)$ represents growth. Note that $P(0) = 2,000$ and $P(1) = 8,000$.

2. C: The number of houses being built each year is 18, and this value is constant. Therefore, the model is increasing at a constant rate each year. The function that best models this scenario is an increasing linear function.

3. D: For each increase in h, the amount of money M increases by $7.45. Therefore, this amount is the hourly wage of the delivery driver.

4. 40: Because the mean is equal to 30, it is true that:

$$\frac{7 + 19 + x + 54}{4} = 30$$

Therefore, $80 + x = 120$, or $x = 40$.

5. B: 208 is the y-intercept of the linear equation. Therefore, when $d = 0$, $H = 208$. If $d = 0$, this means that Chris has not worked any days yet that week. Therefore, the total number of houses that he needs to visit each week is equal to the quantity obtained when plugging $d = 0$ into the equation, resulting in the y-intercept.

Practice Test

Reading

Questions 1–10 are based on the following passage:

Lily had just time to take her seat before the train started, but having arranged herself in the corner with the instinctive feeling for effect which never forsook her, she glanced about with the hope of seeing some other members of the Trenors' party. She wanted to get away from herself, and conversation was the only means of escape that she knew.

Her search was rewarded by the discovery of a very blond young man with a soft reddish beard, who, at the other end of the carriage, appeared to be dissembling himself behind a folded newspaper. Lily's eye brightened, and a faint smile relaxed the drawn lines of her mouth. She had known Mr. Percy Gryce was to be at Bellomont, but she had not counted on the luck of having him to herself on the train; and the fact banished all perturbing thoughts of Mr. Rosedale. Perhaps, after all, the day was to end more favourably than it had begun.

She began to cut the pages of a novel, tranquilly studying her prey through downcast lashes while she organized a method of attack. Something in his attitude of conscious absorption told her that he was aware of her presence: no one had ever been quite so engrossed in an evening paper! She guessed that he was too shy to come up to her and that she would have to devise some means of approach which should not appear to be an advance on her part. It amused her to think that anyone as rich as Mr. Percy Gryce should be shy; but she was gifted with treasures of indulgence for such idiosyncrasies, and besides, his timidity might serve her purpose better than too much assurance. She had the art of giving self-confidence to the embarrassed, but she was not equally sure of being able to embarrass the self-confident.

She waited till the train had emerged from the tunnel and was racing between the ragged edge of the northern suburbs. Then, as it lowered its speed near Yonkers, she rose from her seat and drifted slowly down the carriage. As she passed Mr. Gryce, the train gave a lurch, and he was aware of a slender hand gripping the back of his chair. He rose with a start, his ingenuous face looking as though it had been dipped in crimson; even the reddish tint in his beard seemed to deepen.

The train swayed again, almost flinging Miss Bart into his arms. She steadied herself with a laugh, and drew back; but he was enveloped in the scent of her dress, and his shoulder had felt the fugitive touch.

"Oh, Mr. Gryce, is it you? I'm so sorry—I was trying to find the porter and get some tea."

She held out her hand as the train resumed its level rush, and they stood exchanging a few words in the aisle. Yes—he was going to Bellomont. He had heard she was to be of the party—he blushed again as he admitted it. And was he to be there for a whole week? How delightful!

But at this point one or two belated passengers from the last station forced their way into the carriage, and Lily had to retreat to her seat.

"The chair next to mine is empty—do take it," she said over her shoulder; and Mr. Gryce, with considerable embarrassment, succeeded in effecting an exchange which enabled him to transport him and his bags to her side.

"Ah—and here is the porter, and perhaps we can have some tea."

She signaled to that official, and in a moment, with the ease that seemed to attend the fulfillment of all her wishes, a little table had been set up between the seats, and she had helped Mr. Gryce to bestow his encumbering properties beneath it.

When the tea came, he watched her in silent fascination while her hands flitted above the tray, looking miraculously fine and slender in contrast to the coarse china and lumpy bread. It seemed wonderful to him that any one should perform with such careless ease the difficult task of making tea in public in a lurching train. He would never have dared to order it for himself, lest he should attract the notice of his fellow-passengers; but secure in the shelter of her conspicuousness, he sipped the inky draught with a delicious sense of exhilaration.

Lily, with the flavour of Selden's caravan tea on her lips, had no great fancy to drown it in the railway brew which seemed such nectar to her companion; but rightly judging that one of the charms of tea is the fact of drinking it together, she proceeded to give the last touch to Mr. Gryce's enjoyment by smiling at him across her lifted cup.

Excerpt from <u>The House of Mirth</u> by Edith Wharton, 1905

1. What is Lily's mood on the train?
    a. Lily is bored; she has a novel with her but is not interested in reading it.
    b. Lily is vain; she has attractively positioned herself and wants someone to admire her.
    c. Lily is excited; she is anticipating going to a house party in the country.
    d. Lily is depressed; Mr. Rosedale will not pay attention to her.

2. Why does Lily feel she is lucky to be on the same train as Mr. Percy Gryce?
    a. Lily wants to distract herself from Mr. Rosedale.
    b. Lily wants to have tea, and Mr. Gryce will pay for it.
    c. Lily is glad to see someone of interest to her that she can converse with.
    d. Lily wants to use her gift of giving "self-confidence to the embarrassed."

3. The author, Edith Wharton, says Lily studies her "prey" and plans a "method of attack." Why does the author use such language to describe her heroine's actions?
    a. Wharton uses this metaphor to highlight Lily's confidence with men.
    b. Wharton is trying to show Lily's weakness for conversation.
    c. Wharton wants the reader to see that Lily is evil.
    d. Wharton is emphasizing Lily's boredom by comparing her to a predator.

4. What lets the reader know that the society Lily lives in might be repressive to women?
    a. Lily has developed "instinctive feeling for effect which never forsook her."
    b. Men such as Mr. Rosedale make her want to "get away from herself."
    c. When Lily sees Mr. Gryce is on the train, she must "devise some means of approach which should not appear to be an advance on her part."
    d. Lily has to help Mr. Gryce stow his luggage under the tea table.

5. If the train had not lurched at the exact moment Lily was passing Mr. Gryce's seat, what would she have done?
    a. "Tripped" on her dress and grasped the back of his seat
    b. Kept on going down the carriage, in the hope of a train lurch on her return trip
    c. Exclaimed, "Mr. Gryce, what a pleasure to see you here! Come sit with me."
    d. Asked him about the article he was reading with such absorption

6. What does the fact that Lily is "gifted with treasures of indulgence for such idiosyncrasies," such as the idea of a rich man also being shy, imply about Lily?
    a. It says Lily is a people pleaser.
    b. It hints that Lily secretly disdains rich men.
    c. It proves that Lily has no social skills.
    d. It says Lily's survival depends on her social skills.

7. At the beginning, the narrator leads the reader through the passage from Lily's perspective. Then, readers see Mr. Gryce's perspective. Finally, the narrator ends with Lily's perspective. From whose perspective is the second to last paragraph?
    a. Lily
    b. Mr. Gryce
    c. The train conductor
    d. The porter

8. Mr. Gryce's admiration of Lily's ability to flawlessly make tea on a moving train MOST implies which of the following?
    a. He is in love with Lily.
    b. He is happy to have someone make tea for him.
    c. He is clumsy.
    d. He, as Lily's "prey," has been captured.

9. Which word BEST describes Lily's frame of mind at the end of the passage?
    a. Amused
    b. Resigned
    c. Ambivalent
    d. Satisfied

10. Lily savors "Selden's caravan tea" and would rather not replace that taste with railway tea. This indicates which of the following?
    a. Lily has explicit tastes when it comes to tea.
    b. Lily does not care for Mr. Rosedale's tea.
    c. Lily does not care for railway tea.
    d. Lily does not like Mr. Gryce.

Questions 11–20 are based on the following passage:

Book One: Of the Causes of Improvement in the Productive Powers, of Labour, and of the Order According to Which its Produce is Naturally Distributed Among the Different Ranks of the People

Chapter 1: Of the Division of Labour

The greatest improvement in the productive powers of labour, and the greater part of the skill, dexterity, and judgment with which it is anywhere directed, or applied, seem to have been the effects of the division of labour.

Pin making

To take an example, therefore, from a very trifling manufacture; but one in which the division of labour has been very often taken notice of, the trade of the pin-maker; a workman not educated to this business (which the division of labour has rendered a distinct trade), nor acquainted with the use of the machinery employed in it (to the invention of which the same division of labour has probably given occasion), could scarce, perhaps, with his utmost industry, make one pin in a day, and certainly could not make twenty. But in the way in which this business is now carried on, not only the whole work is a peculiar trade, but it is divided into a number of branches, of which the greater part are likewise peculiar trades. One man draws out the wire, another straights it, a third cuts it, a fourth points it, a fifth grinds it at the top for receiving the head; to make the head requires two or three distinct operations; to put it on is a peculiar business, to whiten the pins is another; it is even a trade by itself to put them into the paper; and the important business of making a pin is, in this manner, divided into about eighteen distinct operations, which, in some manufactories, are all performed by distinct hands, though in others the same man will sometimes perform two or three of them. I have seen a small manufactory of this kind where ten men only were employed, and where some of them consequently performed two or three distinct operations. But though they were very poor, and therefore but indifferently accommodated with the necessary machinery, they could, when they exerted themselves, make among them about twelve pounds of pins in a day. [Smith calculates that] each person ... might be considered as making four thousand eight hundred pins a day. But if they had all wrought separately and independently, and without any of them having been educated to this peculiar business, they certainly could not each of them have made twenty, perhaps not one pin in a day. ...

In every other art and manufacture, the effects of the division of labour are similar to what they are in this very trifling one; though, in many of them, the labour can neither be so much subdivided, nor reduced to so great a simplicity of operation. The division of labour, however, so far as it can be introduced, occasions, in every art, a proportionable increase of the productive powers of labour. The separation of different trades and employments from one another seems to have taken place in consequence of this advantage. This separation, too, is generally carried furthest in those countries which enjoy the highest degree of industry and improvement; what is the work of one man in a rude state of society being generally that of several in an improved one. In every improved society, the farmer is generally nothing but a farmer; the manufacturer, nothing but a manufacturer. The labour, too, which is necessary to produce any one complete manufacture is almost always divided among a great number of hands.

This great increase of the quantity of work which, in consequence of the division of labour, the same number of people are capable of performing, is owing to three different circumstances; first, to the increase of dexterity in every particular workman; secondly, to the saving of the time which is commonly lost in passing from one species of work to another; and lastly, to the invention of a great number of machines which facilitate and abridge labour, and enable one man to do the work of many.

Excerpt from An Inquiry Into the Nature and Causes of the Wealth of Nations by Adam Smith, 1776

11. Which description BEST explains how the division of labor increases the quantity of work in a manufacturing process?
 a. Each worker can efficiently apply their skills and knowledge to specific operations with the least effort.
 b. Work can be passed quickly from one worker to the next.
 c. Workers can more easily share their skills among other workers.
 d. Multiple workers can share their skills by operating a manufacturing machine.

12. Which statement BEST describes the division of labor?
 a. When a group of manufacturing owners and laborers agree that the owners will pay salaries and expenses and sell their products while laborers will make the products
 b. When a group of laborers manufacture an item that requires multiple operations in the process and each laborer with the appropriate skill performs each operation
 c. When manufacturing owners assign the largest number of workers to the operations that require the most complicated operations
 d. When manufacturing laborers are assigned to groups that will each work morning, afternoon, and weekend shifts

13. Which circumstance explains the increased quantity of work when using the division of labor?
 a. It is possible to easily replace a laborer with a machine that performs the same specialized operation with greater efficiency.
 b. All laborers are equally skilled and are best used when they perform multiple types of operations.
 c. Passing work items from one worker to another saves time.
 d. Training each worker to perform multiple operations helps increase their skills and output.

14. How has the division of labor been applied to areas other than manufacturing industries?
 a. To agriculture, in which a farmer performs a multitude of tasks, such as planting, irrigating, and harvesting crops
 b. To nations, each of which exports a specific good, such as coffee or oil
 c. To trades, whose workers provide specialized services, such as banking, retailing, and insurance
 d. To the construction companies that utilize workers who perform all the operations (carpentry, electrical, and plumbing) necessary to complete a project

15. The division of labor would be LEAST appropriate for which of the following type of workers?
    a. Insurance agents
    b. Airline pilots
    c. Software salespeople
    d. Childcare providers

16. Which of the following is NOT an advantage provided by the division of labor?
    a. It reduces time to complete each item.
    b. It produces more items in a given period of time.
    c. It allows workers of any skill level to work independently and concentrate on their work.
    d. It utilizes workers who are not highly skilled to increase their level of productivity.

17. How does the division of labor improve productivity?
    a. Each worker is assigned to an operation that requires the worker's specific skill.
    b. The largest number of workers is assigned to operations that are the easiest to perform.
    c. Workers are equally divided into groups, each of which is assigned to an operation.
    d. The largest number of workers is assigned to operations that are the most complex.

18. Adam Smith notes that the trade of pin making comprises other trades. Which of the following trades could be added to the ones Smith describes?
    a. Pointing the pin
    b. Grinding the pin head
    c. Putting the pins on a card
    d. Assessing the final quality of the pins

19. Which choice would BEST replace "rude" in the following sentence? "This separation, too, is generally carried furthest in those countries which enjoy the highest degree of industry and improvement; what is the work of one man in a rude state of society being generally that of several in an improved one."
    a. Impolite
    b. Crude
    c. Less developed
    d. Benighted

20. Considering the passage as a whole, which of the following BEST describes Adam Smith?
    a. A man who has carefully studied the trade of pin making
    b. A man who knows how to present an argument persuasively
    c. A man who approves of the advances that division of labor has provided
    d. All of the above

Questions 21–30 are based on the following passage:

To the People of the State of New York:

Among the numerous advantages promised by a well-constructed Union, none deserves to be more accurately developed than its tendency to break and control the violence of faction. …

By a faction, I understand a number of citizens, whether amounting to a majority or a minority of the whole, who are united and actuated by some common impulse of passion, or of interest, adversed to the rights of other citizens, or to the permanent and aggregate interests of the community. ...

If a faction consists of less than a majority, relief is supplied by the republican principle, which enables the majority to defeat its sinister views by regular vote: It may clog the administration, it may convulse the society; but it will be unable to execute and mask its violence under the forms of the constitution. When a majority is included in a faction, the form of popular government on the other hand enables it to sacrifice to its ruling passion or interest, both the public good and the rights of other citizens.

The latent causes of faction are thus sown in the nature of man; and we see them every where brought into different degrees of activity, according to the different circumstances of civil society. A zeal for different opinions concerning religion, concerning government, and many other points, as well of speculation as of practice; an attachment to different leaders ambitiously contending for pre-eminence and power; or to persons of other descriptions whose fortunes have been interesting to the human passions, have in turn divided mankind into parties, inflamed them with mutual animosity, and rendered them much more disposed to vex and oppress each other, than to co-operate for their common good. ...

A pure democracy, by which I mean a society consisting of a small number of citizens, who assemble and administer the government in person, can admit of no cure for the mischiefs of faction. A common passion or interest will, in almost every case, be felt by a majority of the whole; a communication and concert result from the form of government itself; and there is nothing to check the inducements to sacrifice the weaker party or an obnoxious individual.

A republic, by which I mean a government in which the scheme of representation takes place, ... promises the cure for which we are seeking. Let us examine the points in which it varies from pure democracy, and we shall comprehend both the nature of the cure and the efficacy which it must derive from the Union. ...

The two great points of difference between a democracy and a republic are: first, the delegation of the government, in the latter, to a small number of citizens elected by the rest; secondly, the greater number of citizens, and greater sphere of country, over which the latter may be extended. ...

The effect of the first difference is, on the one hand, to refine and enlarge the public views, by passing them through the medium of a chosen body of citizens, whose wisdom may best discern the true interest of their country, and whose patriotism and love of justice will be least likely to sacrifice it to temporary or partial considerations. Under such a regulation, it may well happen that the public voice, pronounced by the representatives of the people, will be more consonant to the public good than if pronounced by the people themselves, convened for the purpose. On the other hand, the effect may be inverted. Men of factious tempers, of local prejudices, or of sinister designs, may, by intrigue, by corruption, or by other means, first obtain the suffrages, and then betray the interests, of the people. The question resulting is, whether small or extensive republics are more favorable to the election of proper guardians of the public weal; and it is clearly decided in favor of the latter by two obvious considerations:

In the first place, it is to be remarked that, however small the republic may be, the representatives must be raised to a certain number, in order to guard against the cabals of a few; and that, however large it may be, they must be limited to a certain number, in order to guard against the confusion of a multitude. Hence, the number of representatives in the two cases not being in proportion to that of the two constituents, and being proportionally greater in the small republic, it follows that, if the proportion of fit characters be not less in the large than in the small republic, the former will present a greater option, and consequently a greater probability of a fit choice.

In the next place, as each representative will be chosen by a greater number of citizens in the large than in the small republic, it will be more difficult for unworthy candidates to practice with success the vicious arts by which elections are too often carried; and the suffrages of the people being more free, will be more likely to centre in men who possess the most attractive merit and the most diffusive and established characters.

The other point of difference is, the greater number of citizens and extent of territory which may be brought within the compass of republican than of democratic government; and it is this circumstance principally which renders factious combinations less to be dreaded in the former than in the latter. The smaller the society, the fewer probably will be the distinct parties and interests composing it; the fewer the distinct parties and interests, the more frequently will a majority be found of the same party; and the smaller the number of individuals composing a majority, and the smaller the compass within which they are placed, the more easily will they concert and execute their plans of oppression. Extend the sphere, and you take in a greater variety of parties and interests; you make it less probable that a majority of the whole will have a common motive to invade the rights of other citizens; or if such a common motive exists, it will be more difficult for all who feel it to discover their own strength, and to act in unison with each other.

Hence, it clearly appears, that the same advantage which a republic has over a democracy, in controlling the effects of faction, is enjoyed by a large over a small republic,—is enjoyed by the Union over the States composing it. ...

The influence of factious leaders may kindle a flame within their particular states, but will be unable to spread a general conflagration through the other states: A religious sect, may degenerate into a political faction in a part of the confederacy; but the variety of sects dispersed over the entire face of it, must secure the national councils against any danger from that source: A rage for paper money, for an abolition of debts, for an equal division of property, or for any other improper or wicked project, will be less apt to pervade the whole body of the union, than a particular member of it; in the same proportion as such a malady is more likely to taint a particular county or district, than an entire state. ...

The Federalist Papers: No. 10, by James Madison, 1878

21. Factions can be in either the minority or the majority. Which type of faction in which form of government poses the greatest risk to the public good and the rights of all citizens?
    a. Majority factions in a popular government
    b. Minority factions in a popular government
    c. Majority factions in a centralized government
    d. Minority factions in a centralized government

22. By mitigating the influence of factions, a government will be better able to do which of the following?
    a. Pass laws for the common good instead of any one specific group.
    b. Reduce the amount of public spending.
    c. Raise taxes to create public infrastructure.
    d. Establish a large standing army.

23. Which of the following groups has the potential to form factions?
    a. Merchants
    b. Farmers
    c. Manufacturers
    d. All of the above

24. According to Madison, what is the difference between a republic and a democracy?
    a. In a republic, citizens vote directly to create laws. In a democracy, citizens elect representatives to create laws.
    b. In a republic, citizens elect representatives to create laws. In a democracy, citizens vote directly to create laws.
    c. In a republic, factions elect representatives who create laws. In a democracy, factions vote directly to create laws.
    d. In a republic, factions vote directly to create laws. In a democracy, factions elect representatives who create laws.

25. What does Madison feel is the driving force behind the creation of factions?
    a. A repressive government
    b. The nature of government
    c. The nature of man
    d. The creativity of man

26. Madison felt that the most common reason for the creation of factions was the unequal distribution of property. In Madison's time, a minority of the citizens—those with wealth and who were well educated or professionally skilled—owned much more property than those without such attributes. In addition, property could take various forms, for example, land, businesses, or factories. In this context, what was the issue of protecting citizens from majority factions that Madison had to address?
   a. Government must protect the rights of property owners against the majority of citizens who do not own property.
   b. Government must protect the rights of property owners regardless of the types of property they own. In other words, government must equally protect the rights of numerous business owners and those of less numerous land owners.
   c. Government must protect the rights of all citizens regardless of whether they own property or not and regardless of the type of property they own.
   d. Government must protect the rights of citizens who are wealthy, well educated, or skilled from the majority of citizens who do not possess those attributes.

27. According to Madison, what risk does a republic pose to good government?
   a. Representatives might create laws that will cause taxes to increase.
   b. Citizens might impeach representatives for trivial offenses.
   c. Representatives might create laws contrary to the public interest due to corruption or local prejudice.
   d. Citizens might vote for representatives who do not fulfill their campaign promises.

28. To help prevent the election of corrupt or incompetent representatives, Madison argues in favor of a large republic. Which of the following reasons does Madison give? (Choose two.)
   a. A large number of voters and candidates make it more likely that competent representatives will outnumber the unfit ones.
   b. A large republic makes it more likely that there will be more educated voters and candidates.
   c. A large number of voters are more likely to correctly assess the integrity and merit of a candidate.
   d. A large republic makes it more likely that representatives will vote for laws that help prevent corruption in the government.

29. A large republic helps mitigate the effects of majority factions for which of the following reasons?
   a. A large republic, due to a large number of representatives, will make citizens feel there is no longer a need to create a majority faction.
   b. A large republic will make it less likely for a majority faction to form because minority factions will create more laws that promote the welfare of the republic and cooperation among the citizens.
   c. There will be a greater variety of factions, each with its own interests, making it less likely that a single majority faction will form among them.
   d. A large republic will be more likely to create laws that prohibit the formation of factions.

30. Why does Madison believe that a federal government is better able to control the effects of a majority faction than the states themselves?
   a. A majority faction in one state cannot undermine the rights of other states under a federal system.
   b. A federal government acts as the ultimate majority faction that cannot be overpowered by a state's majority faction.
   c. The laws passed by a state's majority faction are invalid unless all the states approve them.
   d. The president can veto any laws imposed by a state's majority faction.

Questions 31–40 are based on the following passage:

... The North (*United States*), in an unrestrained intercourse with the South, protected by the equal laws of a common government, finds in the productions of the latter great additional resources of maritime and commercial enterprise and precious materials of manufacturing industry. The South, in the same intercourse, benefiting by the agency of the North, sees its agriculture grow and its commerce expand. Turning partly into its own channels the seamen of the North, it finds its particular navigation invigorated; and, while it contributes, in different ways, to nourish and increase the general mass of the national navigation, it looks forward to the protection of a maritime strength, to which itself is unequally adapted. ...

While, then, every part of our country thus feels an immediate and particular interest in union, all the parts combined cannot fail to find in the united mass of means and efforts greater strength, greater resource, proportionably greater security from external danger, a less frequent interruption of their peace by foreign nations; and, what is of inestimable value, they must derive from union an exemption from those broils and wars between themselves, which so frequently afflict neighboring countries not tied together by the same governments, which their own rival ships alone would be sufficient to produce, but which opposite foreign alliances, attachments, and intrigues would stimulate and embitter. Hence, likewise, they will avoid the necessity of those overgrown military establishments which, under any form of government, are inauspicious to liberty, and which are to be regarded as particularly hostile to republican liberty. In this sense it is that your union ought to be considered as a main prop of your liberty, and that the love of the one ought to endear to you the preservation of the other. ...

In contemplating the causes which may disturb our Union, it occurs as matter of serious concern that any ground should have been furnished for characterizing parties by geographical discriminations, Northern and Southern, Atlantic and Western; whence designing men may endeavor to excite a belief that there is a real difference of local interests and views. One of the expedients of party to acquire influence within particular districts is to misrepresent the opinions and aims of other districts. You cannot shield yourselves too much against the jealousies and heartburnings which spring from these misrepresentations; they tend to render alien to each other those who ought to be bound together by fraternal affection.

The alternate domination of one faction over another, sharpened by the spirit of revenge, natural to party dissension, which in different ages and countries has perpetrated the most horrid enormities, is itself a frightful despotism. But this leads at length to a more formal and permanent despotism. The disorders and miseries which result gradually incline the minds of men to seek security and repose in the absolute power of an individual; and sooner or later the chief of some prevailing faction, more able or more fortunate than his competitors, turns this disposition to the purposes of his own elevation, on the ruins of public liberty. ...

It serves always to distract the public councils and enfeeble the public administration. It agitates the community with ill-founded jealousies and false alarms, kindles the animosity of one part against another, foments occasionally riot and insurrection. It opens the door to foreign influence and corruption, which finds a facilitated access to the government itself through the channels of party passions. Thus the policy and the will of one country are subjected to the policy and will of another. ...

The great rule of conduct for us in regard to foreign nations is in extending our commercial relations, to have with them as little political connection as possible. So far as we have already formed engagements, let them be fulfilled with perfect good faith. Here let us stop. Europe has a set of primary interests which to us have none; or a very remote relation. Hence she must be engaged in frequent controversies, the causes of which are essentially foreign to our concerns. Hence, therefore, it must be unwise in us to implicate ourselves by artificial ties in the ordinary vicissitudes of her politics, or the ordinary combinations and collisions of her friendships or enmities.

George Washington's Farewell Address, delivered in 1796

31. Which of the following is a mutual benefit of unrestricted commercial trade between the northern and southern regions of the United States?
    a. A common currency that is recognized in both regions
    b. Accelerated expansion into the western region
    c. A strengthened national maritime presence
    d. Lower tariffs and fewer embargos needed to reduce foreign imports

32. Combining the effort and resources of all Americans from every region helps ensure which of the following?
    a. Greater protection from foreign attacks on the United States
    b. Avoidance of involvement in wars between neighboring nations
    c. A smaller standing army
    d. All of the above

33. The creation of a large military establishment poses a threat to which of the following?
    a. Freedom from involuntary military service
    b. Ample money for public projects
    c. Low taxes
    d. Personal liberty

34. How do political parties gain power among geographical regions?
    a. They promote the belief that there are competing interests and values among regions.
    b. They promise to lower taxes in their own region while arguing that other regions should pay more.
    c. They promote the belief that there are no differences in interests and values among regions.
    d. They encourage voter participation within their regions.

35. In 1812, members of the Democratic–Republican party led by President Madison were able to persuade Congress to declare war against Britain. In New England, the Federal party was in the majority and was anti-war. As a result, two New England states refused to place their state militias under federal control. How did these events support Washington's arguments with regard to national unity and political parties? Which of the following choices are correct?

    I. The establishment of political parties enabled a region within the United States to act against (withhold military forces) the national interest as defined by the U.S. Congress.
    II. The establishment of political parties enabled the one party to declare war in spite of the lack of consent by the majority of citizens in a specific region.
    III. A large standing army made it possible for the federal government to impose martial law in the anti-war New England states.
    IV. The war created enmity between the two political parties, making it more difficult to collaborate on a compromise that was acceptable to all citizens.

    a. Choices I and II
    b. Choices I, II, and III
    c. Choices I, II, and IV
    d. Choices II, III, and IV

36. What did Washington think would happen as a result of the alternate domination of one party over the other?
    a. Unresolved political disputes would drive each political party to pursue absolute power.
    b. Political parties would frequently disband and be replaced by new parties.
    c. Political parties would spend more time running election campaigns.
    d. Political parties would work harder to create a spirit of compromise to settle their disputes.

37. What did Washington think might be the ultimate outcome of a political party that has established domination?
    a. The dominant political party would promote candidates who were unqualified for their positions.
    b. A leader within a dominant party would use their position to gain absolute power rather than act in the public's interest.
    c. Leaders within the dominant and minority party would collaborate to create national programs and policies.
    d. The minority party would adopt the principles and values of the dominant party.

38. What did Washington say about political parties with regard to public councils and the administration of public services?
    a. Political parties weaken them by creating fear and violent rivalry among the public.
    b. Political parties can strengthen them by encouraging political forums and debates.
    c. Political parties can support them by advocating civic harmony by way of parades and speeches.
    d. Political parties can help advocate for more public services.

39. How does the rivalry of political parties affect foreign affairs?
    a. It serves as a model for foreign governments to emulate.
    b. It provides a way for foreign diplomats to learn of the various political viewpoints, which helps them shape their policies with the United States.
    c. It helps foreign diplomats gain direct contact with all the branches of the U.S. government.
    d. It facilitates access to the U.S. government through party operatives, which encourages foreign influences and corruption.

40. What is the primary rule of conduct for the United States in regard to foreign nations?
    a. To expand international commerce and avoid political ties with foreign nations
    b. To develop political ties with foreign nations and expand international commerce
    c. To establish mutual trade protections and form military alliances with specific countries
    d. To avoid political ties with foreign nations and promote tariffs on all imported goods

Questions 41–47 are based on the following passages:

Passage 1: Earth Forces

Since its very beginning, Earth's surface has always been changing. Sometimes the changes happen slowly, and sometimes they happen very rapidly. Sometimes the surface changes are the result of acts of nature, and other times they are the result of man-made efforts. Often, the results appear to be destructive, but sometimes the results are constructive. Two classifications of change are endogenic and exogenic. **Endogenic** (internal) forces work below the ground to change the earth, and **exogenic** (external) forces work above ground to shape the earth.

The following table lists some of the forces of change at work on Earth's surface that could be considered acts of nature.

Force	Description
Earthquake	An endogenic force that causes violent movements of the earth, particularly in the lithosphere
Volcano	A mountain or hill, typically conical in shape, having a rupture, crater, or vent through which lava, rocks, hot vapor, and gas erupt
Tsunami	An unnaturally long, high sea wave; a displacement of water caused by an earthquake, submarine landslide, or other disturbance
Sinkhole	A depression in the land, typically the result of water erosion
Landslide	A ground movement in which a mass of earth or rocks slide from a mountain or cliff
Glacier movement	A slow-moving mass of ice originating from an accumulation of snow

The following are some man-made forces that change Earth's surface.

Force	Description
Dam	A barrier constructed to stop or restrict the flow of water
Land terracing	A term for digging broad channels across the slope of rolling land to reduce runoff, soil erosion, and sediment delivery from higher ground
Levee	A ridge or wall built to regulate water levels, typically to prevent a river from overflowing
Deforestation	The practice of clearing a wide swath or area of trees
Seawall	A man-made structure, typically made of stone or concrete, that extends from the shore into a waterway to mitigate beach erosion
Windbreak	A line of trees or shrubs planted closely together to break the force of the wind

The tables list just a few of the endogenic and exogenic forces constantly at work to change the surface of the earth. In addition, weathering, biological or man-made, breaks down the minerals in rocks and changes their surfaces. Erosion caused by wind, either natural or the result of man-made machines, moves sediment to new locations. Water erosion wears away rocks and also moves sediment, silt, sand, and clay to new locations and can form a delta at the end of a river. Floods, caused by heavy rain or melting snow, also move sediment to different locations. When the waters recede, new sediment takes the place of the old. One of the questions Earth scientists set themselves to answer is: Was this force destructive, constructive, or both?

Passage 2: Earthquakes

In the passage above, the definition of *earthquake* is "an endogenic force that causes violent movements of the earth, particularly in the lithosphere." The initial energy of an endogenic force comes from below the earth. In simpler terms, an earthquake is a violent movement of the earth. An earthquake's force can be measured by a seismograph, which then translates the earthquake's energy level into a magnitude scale between 1 and 10.

A massive earthquake devastated Mexico City on September 19, 1985. The magnitude was 8.1 on the scale of 10. The result was thousands of deaths. Some buildings were reduced to rubble, and others were severely damaged. However, the curious thing about the destruction from this earthquake was that buildings that were medium in height (between six and fifteen stories tall) were the ones that sustained the most wreckage. Taller buildings and shorter buildings were not completely spared, but the damage the earthquake imparted to them was far less severe.

Why would some buildings escape severe damage while others were knocked completely down to their foundations? The answer is the phenomenon of resonance, which is the result of an external force or a vibrating system causing another system close by it to vibrate with greater amplitude. When an earthquake occurs, it releases energy in seismic waves. When the waves hit a building, the lower levels move, and the building begins to vibrate and sway.

All objects, including buildings, have natural frequencies, as do other structures such as bridges. If the frequency of a seismic wave is close to the natural frequency of a building, the amplitude of the oscillations of the building will be increased, and the lower levels of the building will sway farther from side to side with each vibration. If there are no man-made structures in place, such as reinforced concrete, cross braces, and rollers or bearings that isolate the base of the building, the building is in danger of collapse.

To build safer, stronger structures—ones that can best withstand the potential damage of an earthquake—scientists have developed a tool called a **shake table**, a platform that simulates earthquake motion to measure seismic effect on building structures. Early shake tables were moved back and forth by hand and later by pendulums and machines. Today's shake tables are fully computerized to precisely measure earthquake motion and the resulting seismic waves. Shake tables can be small, testing models of buildings that architects have designed, or they can be large enough to hold a full-scale replica of a building that is yet to be built as a permanent structure.

41. In the table listing forces of change at work on Earth's surface that could be considered acts of nature, which of the following is true?
    a. An earthquake occurs slowly, building up strength until its force is released.
    b. The only cause of a tsunami is an earthquake.
    c. Sinkholes develop very quickly.
    d. A landslide is an exogenic force.

42. In the table listing forces of change at work on Earth's surface that could be considered acts of nature, which of the following could be considered endogenic forces?
    a. Earthquake and volcano
    b. Earthquake and glacier movement
    c. Tsunami and glacier movement
    d. Volcano and landslide

43. In the table listing forces that change Earth's surface that are man-made, which of the following are exogenic?
    a. Dam and land terracing
    b. Seawall and windbreak
    c. Levee and seawall
    d. All of the above

44. What would be an appropriate addition to the table that lists destructive forces that change Earth's surface that are man-made?
    a. Drought
    b. Mining
    c. Organic gardening
    d. Avalanches

45. When a flood occurs and the existing sediment is washed away and replaced by different sediment, is this a destructive act?
    a. Yes because the environment has been destroyed
    b. Yes because nothing will grow in the new sediment
    c. No because it hasn't been proven that nothing will grow in the new sediment
    d. It is both destructive and constructive.

46. Reading from context, what does the term *lithosphere* mean?
    a. The surface of the earth
    b. The atmosphere
    c. The South Pole
    d. The core of the earth

47. In the Mexico City earthquake, which buildings sustained the most damage?
    a. The tallest buildings
    b. The medium-height buildings
    c. The ones closest to the city center
    d. The ones closest to where the earthquake hit

Writing and Language

Career-Focused Passage

Train engineers, otherwise known as locomotive engineers, are individuals who drive trains carrying either passengers or cargo. They travel long distances and drive diesel, diesel-electric, or electric trains. The duties of a train engineer are (1) many from ensuring on-time arrivals of passengers or cargo, to monitoring weather and other environmental conditions and maintaining consistent communication with dispatchers at the train's intended destination(s), (2) which is important because different weather conditions require different driving strategies.

To become and remain a train engineer, a person (3) has completed high school, be 21 years of age, and have the following important (4) skills communication, strong hand-eye coordination, problem-solving abilities, and a keen attention to detail. Much of the training a train engineer will encounter will be on-the-job training. (5) Alternatively, they will also have to complete coursework and several exams. These exams (6) includes hearing and vision exams as well as knowledge exams on trains. Many locomotive companies partner with local community colleges or vocational programs to train employees. Furthermore, an (7) engineers performance on the job is periodically tested. Finally, a train engineer must obtain and maintain his or her train engineer license.

(8) Often, before becoming an engineer, an employee may begin their career in the rail industry as a train brakeman, dispatcher, or machinist. A brakeman spends time communicating with other rail professionals, moving cars, and operating switches that allow trains to shift direction. A (9) machinist, on the other hand; works with computer programs related to train operations. (10) Sometimes even writing them.

Advancement opportunities for train engineers include becoming a supervisor, a terminal manager, or a trainmaster. (11) In addition to the aforementioned advancement opportunities for train engineers, it is also possible for train engineers to advance into regulatory, safety inspector, and development roles. A degree may be required for some advanced positions. Alternatively, a similar role within the rail industry is conductor, who supervises cargo and passengers being loaded onto train cars. They also monitor cargo and passengers as they embark on their journey toward their destination. Conductors are often mistaken for those who drive the train, but the engineer completes this task. Both positions are equally important, as are all jobs within the rail industry. Without the engineer, conductor, brakeman, machinist, and supervisors all working in tandem, the locomotive industry could not operate in an efficient manner.

1. Choose the correct answer.
    a. No change
    b. many, from
    c. many; from
    d. many. From

2. The modifying phrase *which is important because different cargo requires different driving strategies* is misplaced. To which place in the first paragraph should it be moved?
    a. It should be the first sentence in the paragraph.
    b. After *electric trains*
    c. After the phrase *from ensuring on-time arrivals of passengers or cargo*
    d. After the phrase *to monitoring weather and other environmental conditions*

3. Choose the correct answer.
    a. No change
    b. must have completed high school
    c. have completed high school
    d. had completed high school

4. Choose the correct answer.
    a. No change
    b. skills; communication
    c. skills: communication
    d. skills...communication

5. Choose the correct answer.
    a. No change
    b. In opposition
    c. In addition
    d. Second

6. Choose the correct answer.
    a. No change
    b. included
    c. include
    d. including

7. Choose the correct answer.
    a. No change
    b. engineer's
    c. engineers'
    d. engineer

8. The writer lists three other jobs in the rail industry (dispatcher, brakeman, and machinist) but only describes two of the jobs. Which information should be added to the author's descriptions?
   a. No additional information needs to be added.
   b. The author should describe in greater detail what it means to be a brakeman.
   c. The author should describe what it means to be a dispatcher.
   d. The author should describe in greater detail what it means to be a machinist.

9. Choose the correct answer.
   a. No change
   b. on the other hand
   c. on the other hand—
   d. on the other hand,

10. The phrase *Sometimes even writing them* is a sentence fragment. What is the best revision for this error?
    a. No change
    b. operations, sometimes even writing them.
    c. Sometimes, they even write the computer programs.
    d. Choices *B* and *C*

11. This writer is concerned that this sentence is too wordy. Which of the following options helps the writer revise for conciseness?
    a. No change
    b. Train engineers can also advance into regulatory, safety inspector, and development roles.
    c. In addition to the aforementioned advancement opportunities, train engineers can also advance into regulatory, safety inspector, and development roles.
    d. It is also possible for train engineers to advance into regulatory, safety inspector, and development roles.

Science-Focused Passage

Disease resistance in plants is a significant factor in a breeder's ability or inability to produce large quantities of market-ready products. Although plants do have natural disease-resistant genes, these are often found in a plant's wild relatives and not plants that have been genetically modified. (12) To aid in the resistance process breeders may introduce their crops to a variety of pathogens to establish resistance. Breeders may also integrate disease-resistant mutated genes found in a plant's wild relatives. This is called *mutation breeding*. Another method of breeding disease-resistant plants is the conventional breeding method. This involves the hybridization of selected plants with disease resistance and the assessment of these hybrids.

Disease resistance, however, is not immediate. (13) The process of establishing go-to-market–ready; disease-resistant products can be a time-consuming and complicated process. However, the other benefit to working so hard to produce disease-resistant plants, in addition to being able to produce larger quantities of food, (14) are that it allows breeders to reduce dependence on lab-created pesticides and fungicides. This in turn reduces the amount of animal breeding done (15) through the use of feed that has been sprayed with chemicals.

As noted above, there are several steps and more than one method a breeder can use (16) to begin the process of producing (17) disease resistant plants through breeding. In mutation breeding, after identifying and introducing existing gene mutations, the breeder will monitor the plants (18), and then determine which are viable prospects for multiplication. (19) Next, the best method for larger-scale production is selected. After this, the plants that have been produced are tested for disease resistance. Finally, the breeders identify those plants with disease resistance. These steps are similar to the steps used in conventional breeding practices.

Steps in producing disease-resistant plants

- Identification and introduction of resistant gene mutations
- Monitoring and selection of prospects for production
- Assessing for disease resistance
- Identification of disease-resistant plants for production

The goal for any breeder is (20) not only producing more food but producing better food. Across the world, entire populations of people lack sufficient nutrients and/or access to quality food. This deficiency can (21) effect both a person's physical and mental health. Through breeding, qualities that can be improved are (22) an increase in vitamins an increase in minerals and an increase in protein. The latter, an increase in protein, can also lessen dependency of agricultural protein producers and unhealthy fat content in some foods. Over the years, hundreds of foods have been produced through both mutation and conventional breeding practices, and hundreds of foods have also been improved. As a result, people throughout the world are living healthier lives.

12. Choose the correct answer.
    a. No change
    b. To aid in the resistance process; breeders
    c. To aid in the resistance process, breeders
    d. To aid in the resistance process breeders,

13. Choose the correct answer.
    a. No change
    b. The process of establishing go-to-market–ready, disease-resistant products
    c. The process of establishing go-to-market–ready disease-resistant products
    d. The process of establishing go-to-market–ready disease-resistant products,

14. Choose the correct answer.
    a. No change
    b. that it allows
    c. allowing
    d. is that it allows

15. Choose the correct answer.
    a. No change
    b. using
    c. with the use of
    d. having the use of

16. Choose the correct answer.
    a. No change
    b. to begin producing
    c. to initiate production of
    d. to produce

17. Choose the correct answer.
    a. No change
    b. Disease Resistant
    c. disease, resistant
    d. disease-resistant

18. Choose the correct answer.
    a. No change
    b. plants and then determine
    c. plants; and then determine
    d. plants. And then determine

19. This sentence uses passive voice. Which of the following can the writer use to revise for active voice?
    a. No change
    b. Next, the breeder selects the best method for larger-scale production.
    c. Next, the best method for larger-scale production is selected by breeders.
    d. Next, the best method is selected for larger-scale production.

20. Choose the correct answer.
    a. No change
    b. not only producing more food but even producing better food
    c. not only producing more food but also producing better food
    d. not only producing more food but more so producing better food

21. Choose the correct answer.
    a. No change
    b. result in
    c. affect
    d. cause

22. Choose the correct answer.
    a. No change
    b. an increase in vitamins, an increase in minerals and an increase in protein
    c. an increase in vitamins. An increase in minerals. And an increase in protein.
    d. an increase in vitamins, an increase in minerals, and an increase in protein

Humanities-Focused Passage

The Impressionism movement found its roots in France in the 1860s. It began when (23) Parisian artists; including Claude Monet and Pierre-August Renoir, gathered together to paint in outdoor settings. This type of art is known as *plein air*. (24) It's purpose, and the purpose of Impressionism, was to paint images that reflected the world in which the artists lived. It was, in essence, a rebellion against classic art, once painted from sketches in a studio. At the time of its inception, however, plein air was anything but accepted by critics. Rather (25) it was compared to wallpaper and dismissed as incomplete.

To the artists, however, painting "en plein air" allowed them to capture light and color in a way not possible in other (26) settings; securing fleeting moments. Impressionist artists would, for example, attempt to paint light as it crossed over water or the effects of shadows on surfaces. Techniques included shorter brush strokes and the separation of colors. (27) As in nature; hard-and-fast outlines were not present, but instead, paintings mimicked realistic environments. Consequently, for many, Impressionist art portrayed more truthfulness than other techniques, encompassing an artist's own impressions of the scene before them.

In the 1870s, despite little support by experts, the artists involved in the Impressionist movement held their first art show. This first show was followed by six more, but by the 1880s, many Impressionists faded away to focus on their individual projects. Impressionism was followed by (28) post-impressionism, morphing its way into other forms of art, including music.

A major contributor to Impressionism in music was a composer (29) by the name of Claude Debussy. Debussy enjoyed the natural sounds and forms music could take apart from the constrictions of traditional music compositions. Like Impressionist painting, Impressionist music veered away from clearly defined structural outlines. Contrasts between melody and harmony become blurred.

As a natural offshoot, in the late 1800s, authors also began to adopt the Impressionist style in their writings. (30) The authors wrote about real life and the real emotions they were experiencing. (31) Descriptions were detailed nor subjective. The goal of most Impressionist authors was to share (32) honest impressions, not impressions that could be validated by a dictionary. One of the most well-known Impressionist authors of her time was Virginia Woolf. As a woman in the late nineteenth century and early twentieth century, she was able to use the Impressionist style of realistic and honest writing to express the female perspective in a world that often restricted a (33) womans' voice. Like the painters and musicians before her, Woolf became a pioneer in her art, paving the way for others who came behind her.

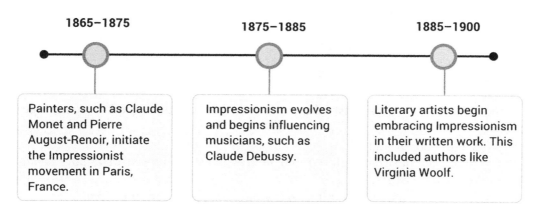

23. Choose the correct answer.
    a. No change
    b. Parisian artists, including Claude Monet and Pierre-August Renoir,
    c. Parisian artists including Claude Monet and Pierre-August Renoir,
    d. Parisian artists. Including Claude Monet and Pierre-August Renoir,

24. Choose the correct answer.
    a. No change
    b. Its' purpose
    c. It is purpose
    d. Its purpose

25. Choose the correct answer.
    a. No change
    b. Rather. It
    c. Rather, it
    d. Rather; it

26. Choose the correct answer.
    a. No change
    b. settings, securing fleeting moments.
    c. setting securing fleeting moments.
    d. settings. Securing fleeting moments.

27. Choose the correct answer.
    a. No change
    b. As in nature. Hard-and-fast
    c. As in nature hard-and-fast
    d. As in nature, hard-and-fast

28. Choose the correct answer.
   a. No change
   b. Post-Impressionism
   c. Post Impressionism
   d. post impressionism

29. Choose the correct answer.
   a. No change
   b. who went by the name of Claude Debussy
   c. named Claude Debussy
   d. denominated Claude Debussy

30. Which of the following sentences best follows the sentence structures used in the passage?
   a. No change
   b. The authors wrote about real life.
   c. Like the painters and musicians before them, Impressionist authors wrote about real life and the real emotions they were experiencing.
   d. Writing about real life and real emotions was what Impressionist authors did.

31. Choose the correct answer.
   a. No change
   b. Descriptions were detailed but subjective.
   c. Descriptions were detailed so subjective.
   d. Descriptions were detailed yet subjective.

32. Choose the correct answer.
   a. No change
   b. honest impressions; not impressions that
   c. honest impressions not impressions that
   d. honest impressions, yet not impressions that

33. Choose the correct answer.
   a. No change
   b. woman's
   c. women's
   d. womans

History-Focused Passage

Deep in the jungles of Guatemala, 300 miles to the north of Guatemala City, lies the mysterious ruins of the ancient city of Holmul. (34) The city stretches for hundreds of miles and is said to shockingly have revealed tens of thousands of timeworn structures. Here, an ancient civilization once thrived (35) (600—700 AD). Millions of inhabitants are said to suddenly have vanished. Even more intriguing (36); the city seems to have been ruled by powerful monarchs known as the Snake Kings. (37) Archaeologist's uncovered still intact tombs, spared from looters and preserved through the centuries. Within these tombs, they found emblems of snakes, leading researchers to believe the serpent symbolized the ancient Mayan kings.

One such tomb was located within a (38) large pyramid, it held the mummy of a man with jade-inlaid teeth. This extravagant bodily embellishment is likely indicative of royalty, as is the hidden placement of the tomb within a deliberately constructed pyramid. Additionally, archaeologists found an inscribed human tibia bone as well as a sun god pendant. Scientists theorize that inscribed bone may be a relic from a prisoner of war. (39) At the point in time when the site around the tomb was being excavated, searchers discovered even more artifacts, including pottery, shells, and an ink pot.

Homul, however, is not the only city that had been led by the (40) snake king dynasty. In a different tomb, a necklace was found with the inscription of a king from a faraway city, one that was situated in what is now present-day Mexico. (41) In fact. Homul is now believed to have been a smaller kingdom that was controlled by larger, more powerful Snake King (42) empires'. These larger empires often warred with rival civilizations and even with their own subjects. This revelation of the existence of multiple, interconnected cities contradicts (43) long held beliefs that the Snake King kingdom functioned reclusively.

It seems, in actuality, that the relationship the Snake King monarchs had with their cities and neighboring civilizations was highly involved and complex. Their strength and commanding (44) presents were immense. This epiphany is why their disappearance is just as surprising and mysterious as their existence. The leading theory behind their sudden departure is that a competing dynasty finally extinguished the Snake King's regime after a longstanding conflict. Other theories include disease, famine, and drought. To this day, researchers continue to sift through the ruins of Homul and surrounding areas in hopes of finally understanding the cryptic reign of the Snake Kings.

34. Which of the following appears in this sentence?
    a. A subject-verb agreement error
    b. An apostrophe error
    c. A lack of appropriate capitalization
    d. A split infinitive

35. Choose the correct answer.
    a. No change
    b. (600:700 AD)
    c. (600–700 AD)
    d. (600–700 "AD")

36. Choose the correct answer.
    a. No change
    b. Even more intriguing the city
    c. Even more intriguing-the city
    d. Even more intriguing, the city

37. Choose the correct answer.
    a. No change
    b. Archaeologists
    c. Archaeologists'
    d. Archaeologist

38. Choose the correct answer.
    a. No change
    b. large pyramid. It held
    c. large pyramid; it held
    d. Choices *B* and *C*

39. The introductory phrase in this sentence is wordy and needs to be revised. Which of the following sentences best revises for conciseness?
    a. When they were searching,
    b. When the site was being dug up,
    c. As the excavation continued,
    d. At that point in time,

40. Choose the correct answer.
    a. No change
    b. Snake king
    c. snake-king
    d. Snake King

41. Choose the correct answer.
    a. No change
    b. In fact: Homul
    c. In fact, Homul
    d. In fact; Homul

42. Choose the correct answer.
    a. No change
    b. empires
    c. empire's
    d. empire

43. Choose the correct answer.
    a. No change
    b. longheld beliefs
    c. long, held beliefs
    d. long-held beliefs

44. Choose the correct answer.
    a. No change
    b. present
    c. presence
    d. presences

Math

No Calculator

1. What is the solution to the following system of linear equations?
$$4x + 6y = 10$$
$$6x - 2y = 12$$

    a. All real numbers

    b. $\left(\frac{23}{11}, \frac{3}{11}\right)$

    c. $\left(\frac{3}{11}, \frac{23}{11}\right)$

    d. No solution

2. What is the solution to the following system of equations?
$$6x - 4y = 14$$
$$y = \frac{3}{2}x + 9$$

    a. (2, 3)
    b. (6, 4)
    c. No solution
    d. All real numbers

3. The local pizza shop is able to make a profit each day if it makes at least $800 in pizza sales. A medium pizza sells for $9 and a large pizza sells for $10. If the shop sells m medium pizzas and l large pizzas, which of the following inequalities represents the number of pizzas needed to be sold to make a profit each day?

    a. $9m + 10l > 800$

    b. $9m + 10l \geq 800$

    c. $\frac{9}{m} + \frac{10}{l} > 800$

    d. $9m + 10l \leq 800$

4. A taxi driver charges $5 per ride and an additional $0.75 per mile. How many miles were driven if a person was charged a fare of $11.75?

5. The cost of producing x newspapers at a publisher is $C(x) = 0.15x + 100$. The publisher sells each newspaper for $2. How many newspapers must be sold to make a profit?

6. What is the solution to the following system of linear equations?
$$8x + 16y = 24$$
$$4x + 8y = 12$$

    a. All real numbers
    b. $\{(x, y) | 4x + 8y = 12\}$
    c. (0, 0)
    d. No solution

7. A zoo charges $10 admission for adults and $8 admission for kids. Last Saturday, 68 people visited the zoo and paid admission. If $640 was collected in admission, which of the following systems of equations could be used to determine how many adults, a, and how many kids, k, visited the zoo:
   a. $10a + 8k = 640$
      $a + k = 68$
   b. $10a + 8k = 680$
      $a + k = 640$
   c. $10a + k = 69$
      $a + 8k = 68$
   d. $a + 8k = 10$
      $10a + k = 8$

8. It costs a shoe manufacturer $40 to assemble 6 pairs of tennis shoes. Also, it costs the same manufacturer $64 to assemble 10 pairs of tennis shoes. Which of the following represents the cost function of manufacturing these tennis shoes, where x represents number of pairs of shoes and C represents cost in dollars?
   a. $C(x) = 6x$
   b. $C(x) = 10x + 6$
   c. $C(x) = 6x + 4$
   d. $C(x) = 6x + 10$

9. A brand of ice cream sells for $4.50 in a 32-ounce container. What is its unit price? Round your answer to the nearest cent.
   a. 7.11 ounces/dollar
   b. $0.14/ounce
   c. $7.11/ounce
   d. 0.14 ounces/dollar

10. Solve the following proportion for x: $\frac{7.6}{x} = \frac{2}{2.1}$.

11. A daycare has a child-to-teacher ratio of 8 to 1. If 128 children are at the daycare, how many teachers would you expect to see at the facility?

12. If a 16-foot tree casts a shadow that is 4 feet long, how long is a shadow cast by a house that is 13 feet tall?

13. 70 percent of what number is 350?

14. A student is drawing a map of the state of Pennsylvania for a homework assignment. The assignment indicates that the map should have a scale factor defining 1 inch on her paper to be 50 miles. Pennsylvania is 283 miles long at its widest point. How many inches on her map would represent this width?
   a. 5 inches
   b. 5 2/3 inches
   c. 14,150 inches
   d. 56 2/3 inches

15. The linear regression model $C = -.75F + 84.3$ is based on nutritional data from 35 packages of protein bars. It describes the relationship between the percent of calories from fat, F, to the percent of calories from carbohydrates, C. Based on this model, which of the following statements must be TRUE?
   a. There is a positive correlation between C and F.
   b. The slope indicates that if F increases by 1, C increases by 0.75.
   c. When 40 percent of the calories in the protein bar are from fat, the percent calories from carbohydrates are predicted to be 54.3.
   d. There is no correlation between C and F.

16. $215 is 18 percent of what amount? Round your answer to the nearest cent if necessary.
   a. $38.70
   b. $1194.44
   c. $1194
   d. $119.44

17. Using the fact that 1 inch is equivalent to 2.54 centimeters, convert 56 feet into centimeters. Round your answer to the nearest tenth if necessary.
   a. 1,706.88 cm
   b. 11.85 cm
   c. 1.84 cm
   d. 22.05 cm

Calculator

18. Justin and Peter both ordered an ice cream cone after their baseball game. The price of Justin's cone was x dollars and the price of Peter's cone was $1 more because he added sprinkles. If Justin and Peter split the cost of their cones and added a 15 percent tip, assuming there is no sales tax, which of the following expressions represents the amount paid for their cones plus tip?
    a. $1.15x + 1$
    b. $1.15x + 1.15$
    c. $2.3x + 1$
    d. $2.3x + 1.15$

19. Which of the following is not a solution to the inequality $9x - 4.5 > -3x + 1.4$?

20. If the graph shown below is reflected over the y-axis and shifted up 5 units, what is the new y-intercept?

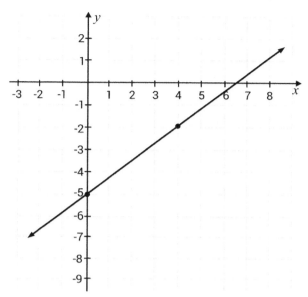

a. (0, −5)
b. (0, 0)
c. (5, 0)
d. (−5, 5)

21. The table below gives the initial population of a species (at time $t = 0$), and the population at each year for 3 years.

Time (years)	Population
0	2,000
1	8,000
2	32,000
3	128,000

Which of the following functions best models the population $P(t)$ at year t?
a. $P(t) = 4{,}000t + 2{,}000$
b. $P(t) = 4{,}000t$
c. $P(t) = 2{,}000(4^t)$
d. $P(t) = 2{,}000(4^{-t})$

22. A housing developer builds twelve new houses a year. Which of the following types of functions best models the number of houses being built as a function of time t.
a. Exponential growth
b. Exponential decay
c. Increasing linear
d. Decreasing linear

23. A pizza delivery driver worked last Saturday night. The amount of money that he made, M, after hour h can be modeled by the equation $M = 7.45h + 16$. What is the best interpretation of the number 7.45 in the context of this problem?
   a. The number of pizzas delivered
   b. The number of hours worked
   c. The total amount of money made on Saturday
   d. The hourly wage of the driver

24. The mean of 7, 19, x and 54 is 30. What is the value of x?

25. Chris works for an internet provider. Each week, he receives a list of houses in which he needs to complete an installation. The number of houses that he needs to visit at the end of each day can be modeled by the equation $H = 208 - 26d$, where H is the number of houses left for installation and d is the number of days that he has worked that week. What is the meaning of the value 208 in this equation?
   a. It represents the number of houses Chris visits each day.
   b. It represents the number of houses Chris must visit each week.
   c. Chris needs 208 days to complete all of his installations.
   d. Chris charges $208 per installation.

26. The following equation shows the height a pole-vaulter reaches as a function of the velocity and the constant, g, of gravity: $h = \frac{v^2}{2g}$. To reach a height of 11 feet, what must the pole-vaulter's velocity be?
   a. 26.5
   b. 5.8
   c. 33.6
   d. 352

27. The dot plots below shows the number of hours students at two schools spent studying for their calculus final exam. Each dot represents 10 students and each school had 200 students taking the calculus final. Which of the following is a correct statement regarding standard deviation for each school?

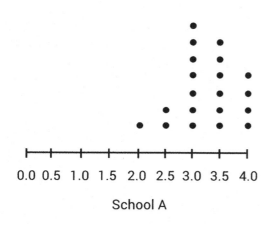

School A

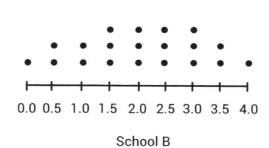

School B

    a. The standard deviation at School B is smaller.
    b. The standard deviation at School B is 0.
    c. The standard deviation at School B is larger.
    d. The standard deviation at School A and B are equal.

28. The given circle has a circumference of 12π. An inscribed angle is 15 degrees. What is the corresponding arc length?
    a. 0.5π
    b. 1.6π
    c. 288π
    d. 0.6π

29. Solve the equation $x^{\frac{2}{3}} = 64$.
    a. 8
    b. 16
    c. 512
    d. 4

30. Choose the expression that is the simplified form of $3x - 3x^2 + 4 - 2x^2$.
    a. $3x + 4 - x^2$
    b. $-5x^2 + 3x + 4$
    c. $-3x^2 + 4 - 2x$
    d. $-5x^2 + x + 4$

31. It takes 45 minutes for Bill to get to the grocery store. Bill spends t minutes riding the bus at an average speed of 0.5 miles per minute. The rest of the time he walks to the store at an average speed of 0.04 miles per minute. Write an equation that represents the total distance traveled.
   a. $d = 0.04t + 0.5(45 - t)$
   b. $d = 0.04t - 0.5(45)$
   c. $d = 0.5t + 0.04(45 - t)$
   d. $d = 0.5t + 0.04t$

32. This month's bank statement says that the balance is $75. You had recorded it as $87. Write an equation to model the amount that you missed recording in your checkbook.
   a. $75 + x = 87$
   b. $87 + 75 = x$
   c. $87 - x = 75$
   d. $75 = 78 - x$

33. Find the missing length of the hypotenuse for a right triangle with side lengths 5 meters and 12 meters.
   a. 11 meters
   b. 13 meters
   c. 7 meters
   d. 17 meters

34. The cost for members and nonmembers to visit the pool at a local gym differs. The members pay an initial fee of $12, and after that they pay $5 a visit. The nonmembers pay a fee of $8 per visit. How many visits will make the initial member fee worthwhile?
   a. 3
   b. 4
   c. 5
   d. 6

35. Darren measured the temperature of the substance he was testing: 53 degrees Fahrenheit. Since he needs to record the temperature in Celsius, he uses the following equation: $F = \frac{9}{5}C + 32$. Rewrite this equation in terms of Fahrenheit so that Darren can convert the temperature.
   a. $C = \frac{5}{9}F + 32$
   b. $C = \frac{9}{5}F - 32$
   c. $C = \frac{5}{9}(F - 32)$
   d. $F = \frac{9}{5}C - 32$

36. The density of a substance is expressed by the formula $d = \frac{m}{v}$, where m is mass and v is volume. A lead sample has a density of 10.4 grams per cubic centimeter and a volume of 1.2 cubic centimeters. What is the mass of the lead sample?
   a. 8.7 grams
   b. 25.0 grams
   c. 12.5 grams
   d. 6.2 grams

37. A linear function is modeled by the equation $f(x) = 4x - 3$. Find the function value of $x = 2$.
    a. 3
    b. 8
    c. -18
    d. 5

38. The new job you have been offered has a starting salary of $40,000 with a 5% raise each year. Write an equation to model the salary after a given number of years.
    a. $s = 40,000 \times 5t$
    b. $s = 40,000 \times 0.05t$
    c. $s = 40,000(0.05)^t$
    d. $s = 40,000(1.05)^t$

39. Simplify the expression $(2x - 3)(3x^2 + x - 4)$.
    a. $6x^3 + 2x^2 + 12$
    b. $2x^3 - 7x^2 - 11x + 12$
    c. $3x^2 + 3x - 7$
    d. $2x^3 + 7x^2 - 3x + 12$

40. Solve the following equation: $6 = \frac{12}{x-4}$.

    a. $\frac{8}{3}$

    b. 2

    c. 6

    d. $\frac{3}{2}$

41. The path of a ball thrown from a hot air balloon is modeled by the equation $h = -16t^2 - 16t + 96$, where h is the height of the ball above the ground and t is the time since the it was thrown. Find the time it takes for the ball to reach the ground.
    a. 2 seconds
    b. 3 seconds
    c. 5 seconds
    d. 12 seconds

42. Solve the equation $y = x^2 + 5x - 14$ using factors and sketch a graph.

a 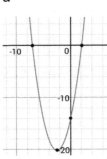
b 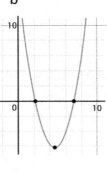
c 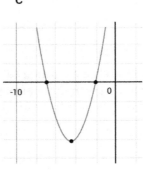
d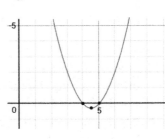

a. x= -7, 2
b. x= 2, 7
c. x= -2, -7
d. x= 5, -14

43. Simplify the polynomial $(z - 4)^3$.
    a. $z^2 - 8z + 16$
    b. $z^3 - 12z^2 + 48z - 64$
    c. $z^3 - 8z^2 + 32z - 64$
    d. $z^3 - 64$

44. Find the product of these complex numbers: $(3 + 2i)$ and $(5 - 3i)$.
    a. $15 - 6i$
    b. $15 - i - 6i$
    c. $21 + i$
    d. $8 + i$

45. Find the sum of the following numbers: $(4 - 2i)$ and $(3 + 5i)$.
    a. $12 - 10i$
    b. $7 + 3i$
    c. $2 + 14i$
    d. $22 + 14i$

46. Find the volume of the following pyramid with a height of 4 cm.

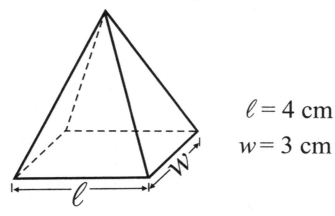

$\ell = 4$ cm

$w = 3$ cm

a. $48\ cm^3$
b. $16\ cm^3$
c. $24\ cm^3$
d. $4\ cm^3$

47. Which variable represents the coordinates equal to $\cos(\frac{3\pi}{4})$?

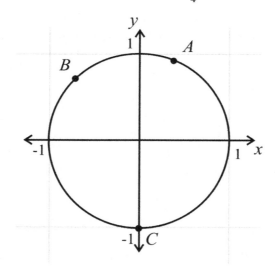

a. A
b. B
c. C
d. None of the above

48. The function notation to represent a cubic relationship is $f(x) = x^3 + 3x$. Find the value of $f(-3)$.
   a. -36
   b. 36
   c. 18
   d. -27

Answer Explanations

Reading

1. C: Lily has an air of excitement about her because she is going to a house party in the country. Lily's excitement is heightened by seeing Mr. Gryce. After she sees him, she begins to formulate a plan to flirt with him, and they spend the train ride sitting together. Although we know that something happened earlier in the day with a Mr. Rosedale that was not pleasant, this does not deter her from enjoying herself on the train.

2. C: Wharton leaves hints when she writes about Lily's "method of attack" and has Lily looking for a porter so she can have tea. Readers might not know what Lily is up to, but we do know that she is glad to see someone she knows, and because of that, she devises a plan to converse with him.

3. A: Wharton uses this metaphor to highlight Lily's confidence with men. Having the confidence of a predator about to pounce on its prey, Wharton is showing that Lily is not afraid to approach Mr. Gryce. Choice B is incorrect; the metaphor does not point to weakness. Choice C is incorrect; we do not know whether Wharton thinks Lily is evil, but the term is too harsh to be applied in this situation. Choice D is incorrect; boredom is not a feeling brought up by comparing Lily to a predator.

4. C: As an unmarried woman, Lily must "devise some means of approach" to an unmarried man. That is, Lily cannot directly approach a single man as a single woman because it is not considered appropriate to do so. Wanting to be perceived as attractive or as someone who will be noticed (Choice A), having unpleasant encounters with other people (Choice B), and being a woman helping a man with luggage (Choice D) do not indicate societal pressure.

5. A: Lily's best option would be "tripping" on her dress. Readers know that Lily is calculating, has her "prey" in sight, and is in attack mode. Choice B is not correct because it shows timidity on her part. She would not go past Mr. Gryce without making a move. Greeting him (Choice C) or asking about what he is reading (Choice D) are ways of meeting him that would be "an advance on her part," which she must avoid.

6. D: In a short passage, readers have been shown the society Lily lives in. She must indulge to survive. Although Choices A and B are true regarding Lily, they are not characteristics demonstrated by that statement. Choice C is not true regarding Lily.

7. B: The second to last paragraph is from Mr. Gryce's perspective. He "watched her in silent fascination while her hands flitted above the tray," which tells us that Mr. Gryce is watching Lily through his own eyes. Thus, it is his perspective we are seeing through here.

8. D: As planned, Lily has beguiled and perhaps captured her prey. Readers don't know if Mr. Gryce loves Lily (Choice A) or is just happy to have tea on the train (Choice B). In this passage, readers have been shown that Mr. Gryce is young, apt to be embarrassed easily, and is also a bit clumsy (Choice C), but his clumsiness is not why he admires Lily's grace.

9. D: Lily is satisfied. She gave self-confidence to the embarrassed, and her day ended on a better note than on which it had begun.

10. D: Lily "had no great fancy" to drown the taste of Seldon's tea, but she did it anyway so as to smile over her cup at Mr. Gryce. Lily also calls the tea "railway brew." This indicates that she does not care for the tea.

11. A: Workers who are skilled can perform their operation quickly. Choice *B* is incorrect because passing work is time consuming and inefficient. Choice *C* is incorrect because each worker has a specialized skill that is best applied by the workers themselves. Choice *D* is incorrect because a single worker can perform the operation of multiple workers using a machine. There is no need to use multiple workers to perform the same operation on the same machine.

12. B: Assigning laborers to operations according to the laborers' skills results in the greatest efficiency. Choice *A* is incorrect because the division of labor applies only to manufacturing laborers. They perform specific operations according to their skills. Choice *C* is incorrect because the division of labor assigns laborers according to their skills and not according to the number of laborers needed for each operation. Choice *D* is incorrect because the division of labor assigns laborers according to their skill and not according to when they are scheduled to work.

13. A: Machines can be used to perform operations that help skilled workers perform their specific tasks or can perform the tasks entirely. Choice *B* is incorrect because each laborer usually has a skill they are best at and can perform most efficiently. Choice *C* is incorrect because passing work between workers is time consuming. Choice *D* is incorrect because a worker is most efficient when they perform the skill in which they are most adept.

14. C: Each trade requires workers with the correct skills to efficiently perform their tasks. Choice *A* is incorrect because a farmer who works on all their tasks is not dividing the labor of planting, irrigation, and harvesting among skilled laborers. Choice *B* is incorrect because the division of labor pertains only to assigning workers to tasks based on their special skills and not the specialization of goods or services. Choice *D* is incorrect because each worker has all the skills necessary to complete all the operations. There is no division of labor.

15. D: Typically, childcare providers perform all the required tasks without relying on the labor of others. Choice *A* is incorrect because other workers, such as insurance underwriters and appraisers, must perform their tasks before a policy can be written. Choice *B* is incorrect because pilots rely on the specialized skills of other workers, such as a navigator, co-pilot, and flight attendants, to perform their tasks in order to fly an airplane. Choice *C* is incorrect because before the software is ready for sale, the software code, user interface, and documentation must be developed by specialized workers.

16. C: Workers lacking the appropriate skill and working alone will do poor work over a long period of time. Choice *A* is incorrect because a worker skilled at a specific operation can perform it quickly and efficiently. Choice *B* is incorrect because collectively, when all workers are working at maximum efficiency, more items will be produced in a given time period. Choice *D* is incorrect because even poorly skilled workers, when sharing a limited number of operations with other workers, are more efficient than when they work individually.

17. A: A worker who is most skilled at a specific operation will be the most productive. Choices *B, C,* and *D* are incorrect because worker skills are not being matched to operations that require specific skills.

18. D: This trade (quality control) is not included in Smith's description but could be added as an additional operation. Choices *A, B,* and *C* are incorrect because they are included in Smith's description.

19. C: Smith is comparing states of societal development in terms of industrialization. Although Choice *A* is a synonym for "rude," Smith is not interested in how polite or impolite the societies are. Choice *B,* "crude," overstates the point Smith is making, and there is no implication that the society is "benighted" (Choice *D*).

20. D: Choices *A, B,* and *C* are all equally true of Smith. The depth of the discussion of pin making proves he researched the topic fully, his argument is presented logically and efficiently, and his enthusiasm for the topic is clear.

21. A: In a popular government, a majority faction can outvote the minority. Choice *B* is incorrect because in a popular government a minority faction cannot outvote the majority. Choices *C* and *D* are incorrect because all power is held by a single authority that may act on behalf of either a majority or minority faction.

22. A: Factions, by definition, act on their own behalf. Mitigating their impact makes it more likely that a government will be able to pass laws that are for the common welfare. Choices *B, C,* and *D* are incorrect because reducing public spending, raising taxes, or establishing a large standing army may be either for the common good or may benefit specific factions.

23. D: Merchants (Choice *A*), farmers (Choice *B*), and manufacturers (Choice *C*) all have interests that are specific to their group. For example, merchants may want to eliminate import duties, thus lowering prices and increasing demand. Farmers may want the government to guarantee a minimum price for a commodity, such as corn or wheat, thus making their products profitable regardless of crop yields. Manufacturers may want to impose import duties to make foreign products more expensive, thus increasing demand for domestic products.

24. B: In a republic, citizens elect representatives to create laws. In a democracy, citizens represent themselves when creating laws. Choice *A,* the opposite, is incorrect. Choices *C* and *D* are incorrect because in a republic or a democracy, citizens either elect representatives or participate directly, respectively. In either case, factions do not define a democracy or a republic but may be created in either form of government.

25. C: Madison believed that individuals are naturally drawn to personal interests, which inclines individuals to join together in a quest for dominance over the minority rather than cooperation for their common good. Choices *A* and *B* are incorrect because the citizens, not the government, create factions. Choice *D* is incorrect because it is not the human creativity but its natural quest for dominance that drives mankind to create factions.

26. C: The government must protect both the rights of property owners and the majority of people who do not own property as well as the rights of property owners regardless of the type of property they own. Choice *A* is incorrect because the government must also protect the rights of property owners regardless of the type of property they own. Choice *B* is incorrect because the government must also protect the rights of all property owners from the majority of people who do not own property. Choice *D* is incorrect because the issue of whether a citizen has wealth, education, or special skills is not relevant to the issue of property ownership.

27. C: A representative, once elected, may choose to act out of self-interest rather than on behalf of the community's best interests. Choice *A* is incorrect because a law that means higher taxes may still be in the community's best interest. Choice *B* is incorrect because it was not stated as a risk by Madison. Choice *D* is incorrect because a representative may not be able to fulfill a campaign promise due to circumstances out of their control.

28. A, C: Madison argues that a large republic will help eliminate incompetent representatives because a large number of voters and candidates make it more likely that competent representatives will outnumber the unfit ones and correctly assess the integrity and merit of a candidate. Choice *B* is incorrect because Madison does not state that the level of education among voters and candidates is a factor. Choice *D* is incorrect because there is no assurance that representatives in either a large or small republic will vote for laws that help prevent corruption.

29. C: A large republic will encompass large regions, each with its own factions, making it less likely that a majority faction could encompass the interests of all the regions. Choice *A* is incorrect because Madison states that it is in the nature of man to pursue his own interests at the expense of the rights of others. Madison does not state that a republic of any size will eliminate the impulse for citizens to gain power by way of a majority faction. Choice *B* is incorrect because factions, by definition, act in their own interest and not in the interest of other factions. Choice *D* is incorrect because such laws would be impractical and violate the liberty of citizens to form factions.

30. A: A state's majority faction does not have the power to impose its laws on other states in a federal government. Choice *B* is incorrect because Madison does not describe the federal government as a majority faction. Choice *C* is incorrect because a state's majority faction may pass its own laws without the approval of other states. Choice *D* is incorrect because Madison does not state that as possible.

31. C: Increased overseas commerce derived from northern manufacturing and southern agriculture will lead to the need for more ships and thus a strengthening of the nation's maritime presence. Choice *A* is incorrect because although a common national currency may make commerce easier, it is not a benefit derived from unrestricted trade. Choice *B* is incorrect because, in this context, the benefits of unrestricted trade between the North and South are not related to western expansion. Choice *D* is incorrect because tariffs and embargos are not necessarily related to unrestricted trade between the North and South.

32. D: All of the choices are correct. Choice A is correct because a unified nation can pose a greater threat to a foreign nation that is considering an attack on the United States. Choice B is correct because a unified nation will be less likely to be drawn into wars that are not in the national interest. Choice C is correct because a nation disinclined to war will not need a large army always ready for war.

33. D: A large military establishment is stated as being hostile to liberty, especially under a republican form of government. Choice A is incorrect because a large military may initiate involuntary military service, but it is not stated as being a threat. Choice B is incorrect because a large military may mean less money to spend on public projects, but it is not stated as being a threat. Choice C is incorrect because taxes may have to be raised to pay for the military, but it is not stated as being a threat.

34. A: In order for parties to gain power in their own districts, they magnify and misrepresent the interests and values of other regions. Choice B is incorrect because although parties may express differences on how taxes are raised, it is not stated as a way of gaining power. Choice C is incorrect because parties promote the differences, not the common interests. Choice D is incorrect because party members may be motivated to vote, but it is not stated as a reason.

35. C: The two parties diverged sharply on the issue of declaring war, thus causing a rift between citizens in specific states and the nation as a whole. According to Washington, if citizens thought of themselves as Americans rather than members of regionally aligned political parties, it would be more likely that a political consensus on the issue of war would have been reached without regional conflict (I and IV). Also, if war had been declared by general consensus, all state militias would have been deployed without issue (I). Choice A is incorrect because it does not include the fact that friction between the two political parties made it difficult to collaborate on a compromise acceptable to everyone (IV). Choices B and D are incorrect because the question describes anti-war states withholding military assistance but not the imposition of martial law due to a large standing federal army (III).

36. A: Washington believed that political parties driven by revenge and party dissension will vie for permanent dominance over other parties. Choice B is incorrect because political parties will seek to gain strength and power and not drop out. Choice C is incorrect because running election campaigns may help gain dominance for an election cycle but does not ensure enduring absolute power. Choice D is incorrect because Washington believed that the natural inclination of those in power is to ensure total domination and not a desire to share power.

37. B: Washington believed the dominant party could maintain control more easily by investing absolute power in an individual. Choice A is incorrect because the unqualified candidates are less likely to be elected and thus maintain dominance. Choice C is incorrect because a dominant party will use its absolute power to promote its own policies and programs. Choice D is incorrect because Washington spoke of the potential for despotism arising from political parties and not on party tactics.

38. A: It is in the self-interest of political parties to provoke animosity toward rival parties and disregard services to the public at large. Choice *B* is incorrect because political parties engender party loyalty by fueling mistrust of rival parties and not encouraging the open exchange of political ideas. Choice *C* is incorrect because political parties do not gain strength by encouraging civic harmony in any respect but rather by fueling mistrust of rival parties. Choice *D* is incorrect because political parties are mainly driven by the need to gain political power and not by public services.

39. D: Each political party is vying with the other for special favors from foreign nations, which leads to conflicting policy issues among countries. Choice *A* is incorrect because foreign governments are not likely to modify their established systems based on their observations of the American system. Choice *B* is incorrect because although foreign governments may gain insight by observing various political viewpoints, the rivalry between parties may directly influence foreign policies in negative ways. Choice *C* is incorrect because the goal of rival political parties is not necessarily to enable efficient communication with all branches of the U.S. government.

40. A: The country should both avoid political ties so as to avoid foreign entanglements that are not in the national interest and promote national wealth by expanding international commerce. Choice *B* is incorrect because the country should avoid political ties so as to avoid foreign entanglements that are not in the national interest. Choice *C* is incorrect because political ties with foreign nations should be avoided and international commerce should be expanded. Military alliances and trade protections with specific nations would both compromise national security and hinder international commerce. Choice *D* is incorrect because import tariffs would restrict rather than expand international commerce.

41. D: A landslide is an exogenic force. Choice *A* is incorrect because earthquakes are rapid occurrences. Choice *B* is incorrect because the passage points out that not only can earthquakes cause tsunamis, but other occurrences such as submarine landslides can cause tsunamis as well. Choice *C* is incorrect because sinkholes are the result of a gradual wearing away of materials below the ground.

42. A: Earthquakes and volcanoes are both internal forces that work below the ground. Choices *B, C,* and *D* all include one exogenic force (glacier movement and landslide).

43. D: All of the choices—Choices *A* (dam and land terracing), *B* (seawall and windbreak), and *C* (levee and seawall)—are exogenic forces.

44. B: Mining is a man-made force that change's the earth's surface. Drought (Choice *A*) and avalanches (Choice *D*) are natural destructive forces but are not man-made. Organic gardening (Choice *C*) is man-made but is a constructive rather than destructive activity.

45. D: Although a flood washes away existing sediment (destructive), it replaces it with new sediment (constructive).

46. A: The passage is about earthquakes, which originate below the earth's surface. Choices *B* (atmosphere), *C* (South Pole), and *D* (core of the earth) are incorrect.

47. B: Even though logic might dictate that the tallest buildings would be most vulnerable, the passage clearly states why the medium-height buildings were the ones to sustain the most damage.

Writing and Language

Career-Focused Passage

1. B: Choice B is correct because it uses a comma to separate an independent and dependent clause. Choice A is incorrect because a comma is necessary between the words *many* and *from*. Choice C is incorrect because a semicolon should only be used to separate two independent clauses, not an independent and a dependent clause. Choice D is incorrect because a period should not be used to separate an independent and a dependent clause.

2. D: Choice D is correct because the phrase *which is important because different cargo requires different driving strategies* is a misplaced modifier phrase. It modifies the phrase *to monitoring weather and other environmental conditions*, but because it is placed at the end of the sentence, it appears to be modifying *destinations*. Choice A is incorrect and would be confusing if it was the first sentence in the first paragraph because the author has not yet discussed driving in different types of weather. Choice B is incorrect because monitoring weather conditions and changing driving strategies does not modify the type of train being driven. Choice C is incorrect because monitoring weather conditions and changing driving strategies does not modify the on-time arrival of passengers and cargo.

3. B: The choice "must have completed high school" is the correct answer choice. The word "must" signifies a requirement, which is in context with the train engineer's duties. Additionally, saying "must have" adds parallel structure to the rest of the list: "must have completed high school . . . [must] be 21 years of age, and [must] have the following important skills."

4. C: Choice C is correct because it uses a colon to introduce a list. Choice A is incorrect because it does not introduce the list with a colon or revise the sentence structure to introduce the list. Choice B is incorrect because it uses a semicolon to introduce a list. A semicolon must only be used to separate independent clauses or to separate items in a list that already contains commas. Choice D is incorrect because it uses an ellipsis to introduce a list. An ellipsis should only be used when words have been omitted from a sentence.

5. C: Choice C is correct because it uses a transition phrase that logically connects the idea in the preceding sentence. In the preceding sentence, the writer tells readers about the training engineers must undergo, and the next sentence provides more information. Thus, *in addition* helps portray this relationship between ideas. Choice A is incorrect because the transition *alternatively* does not logically show the relationship between the two sentences. Choice B is incorrect because the relationship between this sentence and the preceding sentence is not one of opposition. Rather, it shows additional information. Choice D is incorrect because a writer cannot use transitions such as *second* or *third* without also using the transition *first*.

6. C: Choice C is correct because the verb *include* agrees in number with the subject *exams*. Choice A is incorrect because *exams* and *includes* do not agree in number. Choice B is incorrect because the verb *included* does not agree in tense with the rest of the passage. Choice D is incorrect because the verb *including* does not agree grammatically with the subject *exams*.

7. B: Choice B is correct because it uses an apostrophe to indicate possession by an individual engineer. Choice A is incorrect because an apostrophe is needed to indicate possession. Choice C is incorrect because it places the apostrophe after the *s* in a singular subject. Choice D is incorrect because the word *engineer* needs an *'s* to indicate singular possession.

8. C: Choice *C* is correct because the job dispatcher is listed but not described. Choice *A* is incorrect because all three jobs should be described. Choice *B* is incorrect because a description of the brakeman job is adequately described. Choice *D* is incorrect because a description of the machinist job is adequately described.

9. D: Choice *D* is correct because it uses a comma to offset the phrase *on the other hand*. Choice *A* is incorrect because a semicolon cannot be used to separate two dependent clauses. Choice *B* is incorrect because a comma is needed after the word *hand*. Choice *C* is incorrect because an em-dash is used to offset the phrase *on the other hand*. Although an em-dash can be used offset non-restrictive phrases such as this, one should appear on each side of the clause, not just at the end.

10. D: Choice *D* is correct because a fragment can be added to another sentence (Choice *B*) or formed into a complete sentence (Choice *C*). Choice *A* is incorrect because a sentence fragment is a grammatical error and needs to be revised.

11. B: Choice *B* is correct because it eliminates unnecessary phrases, such as *In addition to the aforementioned advancement opportunities* and *it is also a possibility*, which can bury a writer's ideas by confusing readers with wordiness. Choice *A* is incorrect because the sentence is wordy and needs to be revised. Choice *C* is incorrect because it retains the wordy phrase, *In addition to the aforementioned advancement opportunities*. Choice *D* is incorrect because it retains the wordy phrase, *it is also a possibility*.

Science-Focused Passage

12. C: Choice *C* is correct because a comma is used to separate an introductory clause from the main clause. Choice *A* is incorrect because a comma is necessary to separate an introductory clause from a main clause. Choice *B* is incorrect because a semicolon should not be used to separate an introductory clause from a main clause. Choice *D* is incorrect because the comma should be placed after *process*, not *breeders*.

13. B: Choice *B* is correct because a comma separates two corresponding adjectives with a comma. Choice *A* is incorrect because two corresponding adjectives should not be separated by a semicolon unless the adjectives are in a list and could confuse readers if separated by commas. Choice *C* is incorrect because a comma must be used to separate two corresponding adjectives to create clarity. Choice *D* is incorrect because the comma should be placed between, not after, the two corresponding adjectives.

14. D: Choice *D* is correct because it uses the verb tense *is*, which matches the singular subject, *benefit*. Choice *A* is incorrect because the subject and verb do not agree. Choice *B* is incorrect because it is missing the linking verb, *is*. Choice *C* is incorrect because the gerund verb form, *allowing*, does not match the tense of the rest of the sentence.

15. B: Choice *B* is correct because it eliminates the wordy phrase, *through the use of*. Choice *A* is incorrect because the phrase *through the use of* is wordy. Choice *C* is incorrect because the phrase *with the use of* is unnecessarily wordy. Choice *D* is incorrect because the phrase *having the use of* is unnecessarily wordy.

16. D: Choice *D* is correct because it eliminates the wordy phrase, *to begin the process of producing*. Choice *A* is incorrect because it retains the wordy phrase, *to begin the process of producing*. Choice *B* is incorrect because *to begin producing* can be consolidated even further by removing *begin* and replacing *producing* with *produce*. Choice C is incorrect because it revises the original phrase, *to begin the process of producing,* to another wordy phrase, *to initiate production of*.

17. D: Choice D is correct because to be unified, the compound adjective *disease-resistant* should remain hyphenated. Choice A is incorrect because the compound adjective *disease-resistant* needs to be hyphenated. Choice B is incorrect because the compound adjective *disease-resistant* is not a proper noun and therefore does not require capitalization; it is also not hyphenated in the answer choice. Choice C is incorrect because compound adjectives should be hyphenated, not separated with a comma.

18. B: Choice B is correct because when using a comma and a coordinating conjunction, the words on each side should be part of independent clauses. Choice A is incorrect because a comma and a coordinating conjunction should not precede a dependent clause. Choice C is incorrect because a semicolon should not separate an independent and a dependent clause. Choice D is incorrect because the clause *and then determine which are viable prospects for multiplication* is dependent and cannot stand alone as a sentence.

19. B: Choice B is correct because the writer uses active voice. Choice A is incorrect because it retains the passive voice. Choice C is incorrect because adding *breeders* as an object of the verb following *selected by* retains the passive voice. Choice D is incorrect because it simply moves the passive verb phrase *is selected* to another place in the sentence.

20. C: Choice C is correct because *not only* should be paired with *but also*. Choice A is incorrect because the conjunction *but* is not followed by *also*. Choice B is incorrect because *not only* should be paired with *but also* instead of *but even*. Choice D is incorrect because *more so* indicates additional, not equal, information. It is also unclear and awkward.

21. C: Choice C is correct because *affect* is a verb that means to impact something. Choice A is incorrect because *effect* is a noun that means result, and a person or thing cannot result another person or thing. Choice B is incorrect because *result in* is grammatically incorrect and also does not logically fit into the context of the sentence. Choice D is incorrect because *cause* is grammatically incorrect.

22. D: Choice D is correct because a comma is placed after each item in the list as well as before the last item. Choice A is incorrect because commas are missing from the list. Choice B is incorrect because a comma is missing before the last item in the list. Choice C is incorrect because each item in the list is a fragment and cannot be independent sentences.

Humanities-Focused Passage

23. B: Choice B is correct because it offsets the phrase *including Claude Monet and Pierre-August Renoir* with commas. Choice A is incorrect because a semicolon should only be used to separate independent clauses or lists with several commas. Choice C is incorrect because a comma is missing before the word *including*. Choice D is incorrect because the phrase ending with *Parisian artists* and the phrase beginning with *including* are both dependent clauses and should be separated with a comma, not a period, and *including* should not be capitalized.

24. D: Choice D is correct because an apostrophe should not be used when showing possession through the word *its*. Choice A is incorrect because the placement of the apostrophe in the word *it's* makes it a contraction for *it is,* which is not the correct meaning in this sentence. Choice B is incorrect because no apostrophe is needed after *its* to indicates possession. Choice C is incorrect because *it is* does not indicate possession.

25. C: Choice *C* is correct because the transition word *rather* is followed by a comma. Choice *A* is incorrect because a comma is missing after the word *rather*. Choice *B* is incorrect because the word *rather* is not a complete sentence and should not be followed by a period or capitalized. Choice *D* is incorrect because *rather* is not an independent clause and should not be followed by a semicolon.

26. B: Choice *B* is correct because the independent and dependent clause in the sentence are separated by a comma. Choice *A* is incorrect because a fragment should not be offset by a semicolon. Choice *C* is incorrect because the independent and dependent clause are not separated by a comma. Choice *D* is incorrect because the dependent clause cannot stand alone as a sentence.

27. D: Choice *D* is correct because the introductory clause is followed by a comma. Choice *A* is incorrect because the introductory clause is followed by a semicolon. Choice *B* is incorrect because the introductory clause is followed by a period. Choice *C* is incorrect because there should be a comma after the introductory clause.

28. B: Choice *B* is correct because the term *Post-Impressionism* is a proper noun, requiring both capitalization and hyphenation. Choice *A* is incorrect because the term *Post-Impressionism* is a proper noun and should be capitalized. Choice *C* is incorrect because the term *Post-Impressionism* requires hyphenation. Choice *D* is incorrect because the term *Post-Impressionism* lacks capitalization and hyphenation.

29. C: Choice *C* is correct because it condenses the wordy phrase *by the name of* to *named*. Choice *A* is incorrect because the current sentence structure retains the wordy phrase *by the name of*. Choice *B* is incorrect because it retains a wordy phrase instead of revising it for conciseness. Choice *D* is incorrect because it contains a complex word instead of a more straightforward word.

30. C: Choice *C* is correct because it matches the complex sentence structures found throughout the passage. Choice *A* is incorrect because it retains a sentence structure that does not flow with the rest of the passage. Choice *B* is incorrect because it revises the sentence into a simple sentence structure, which does not flow with the rest of the passage. Choice *D* is incorrect because it uses passive voice, which differs from the active voice used throughout the rest of the passage.

31. D: Choice *D* is correct because it uses the most appropriate coordinating conjunction *yet* to separate ideas. Choice *A* is incorrect because it retains the awkward and out-of-place conjunction *nor* to separate ideas. Choice *B* is incorrect because it uses the conjunction *but*, which reflects opposition; however, no opposition is expressed in this sentence. Choice *C* is incorrect because it uses the conjunction *so*, which indicates a cause-and-effect relationship, but no cause-and-effect relationship is expressed in this sentence.

32. A: Choice *A* is correct because the comma is appropriately placed before the dependent clause in the sentence. Choice *B* is incorrect because a semicolon, not a comma, is used before the non-restricted, dependent clause. Choice *C* is incorrect because a comma should appear before the non-restricted, dependent clause. Choice *D* is incorrect because it uses the conjunction *yet*, meaning nonetheless, which does not fit within the context of this sentence.

33. B: Choice *B* is correct because it places an apostrophe before the *s* in the singular possessive noun, *woman's*. Choice *A* is incorrect because the noun *woman* is singular, not plural, and thus, the apostrophe is incorrectly placed after the *s*. Choice *C* is incorrect because *women* is plural. Choice *D* is incorrect because it lacks an apostrophe to show possession.

History-Focused Passage

34. D: Choice D is correct because the sentence includes the split infinitive *to shockingly have*. The main verb, *have*, is separated from *to* by *shockingly*. Choice A is incorrect because the subjects and verbs in this sentence all agree in number, meaning plural and singular verbs match. Choice B is incorrect because there is no possession in this sentence, nor are there any contractions, so an apostrophe is not needed. Choice C is incorrect because the only word in this sentence requiring capitalization is the first word of the sentence, and it is capitalized.

35. C: Choice C is correct because an en-dash is used to show a time duration between 600 and 700 AD. Choice A is incorrect because it uses an em-dash instead of an en-dash. Choice B is incorrect because a colon is used to show a passage of time. A colon should only be used to introduce a list or provide additional information to a main clause. Choice D is incorrect because it places quotation marks around AD. Quotation marks should be used around other people's words or titles of short works. AD is part of the time period being referenced.

36. D: Choice D is correct because it uses a comma after the introductory phrase, *Even more intriguing*. Choice A is incorrect because it uses a semicolon. Choice B is incorrect because there is no punctuation separating the introductory clause from the rest of the sentence. Choice C is incorrect because it uses a hyphen to separate the introductory phrase from the main part of the sentence. The function of a hyphen is to join words or parts of words, not sentences or parts of sentences.

37. B: Choice B is correct because the word *archaeologists* functions as a noun acting on the verb *uncovered*. It is not possessive, and it is not a contraction, which are the two instances that would require an apostrophe. Choice A is incorrect because the word *archaeologists* retains an apostrophe. Choice C is incorrect because the use of the apostrophe in *archaeologists'* indicates possession, but there is no possession in this sentence. Choice D is incorrect because *archaeologist* is singular rather than plural.

38. D: Choice D is correct because two independent clauses can be separated by a period (Choice B) or a semicolon (Choice C). If independent clauses are separated with a period, the first word of each sentence should be capitalized. Choice A is incorrect because two independent clauses cannot be separated by a comma.

39. C: Choice C is correct because it uses the fewest words to make the author's point, and it is specific. It identifies the activity of the excavation and points out that these additional findings took place after the initial tomb was discovered. Choice A is not correct because it uses a vague pronoun, *they*, and does not clearly identify what is being searched and when. Choice B is incorrect because it does not clearly identify which site is being dug up and when. In addition, the words *dug up* can be consolidated even further with words such as *searched* or *excavated*. Choice D is incorrect because it uses a vague and wordy phrase, *at that point in time*. Readers are not told when this time is, and phrases such as *at that point in time* can be consolidated even further to words such as *when* and *at the time*.

40. D: Choice D is correct because it capitalizes the proper noun, *Snake King*. Choice A is incorrect because it does not capitalize the proper noun, *Snake King*. Choice B is incorrect because it only capitalizes the word *Snake*. Choice C is incorrect because the term *Snake King* is not hyphenated.

41. C: Choice C is correct because it uses a comma to offset the introductory phrase *In fact*, which helps a writer clearly distinguish ideas and maintain flow. Choice A is incorrect because a period should not be used to offset an introductory clause from the main clause of a sentence. Furthermore, the phrase *In* *fact* is not a complete sentence and cannot stand alone. Choice B is incorrect because a colon should not be used to separate an introductory clause from a main clause. Choice D is incorrect because a semicolon should not be used to separate an introductory clause from a main clause.

42. B: Choice B is correct because it eliminates the apostrophe from the word *empires*, which does not indicate possession in this sentence. Choice A is incorrect because it retains an apostrophe in the word *empires*. Choice C is incorrect because it uses an apostrophe in the word *empires*. Choice D is incorrect because it uses *empire*, a singular noun, but the author is referring to multiple *empires*, requiring the plural noun.

43. D: Choice D is correct because it uses a hyphen to join the words *long* and *held*, which forms a compound modifier *long-held,* that modifies the word *beliefs*. Choice A is incorrect because it does not use a hyphen to join the compound modifier *long-held*. Choice B is incorrect because it joins the words *long* and *held* without a hyphen. Choice C is incorrect because it joins the words *long* and *held* with a comma rather than a hyphen.

44. C: Choice C is correct because in this sentence, the author is speaking of the Snake Kings' existence (i.e., their *presence*). Choice A and Choice B are incorrect because a *present* is a gift, and *presents* and *present* are the plural and singular forms of this noun, respectively. Choice D is incorrect because it uses the word's plural form, *presences,* instead of the singular form.

Math

No Calculator

1. B: This system is best suited for elimination because both equations are in standard form. Multiply the second equation times 3. This step results in the equivalent equations $18x - 6y = 36$. Now, the coefficients on the y −variable are opposite. Add the equations together to obtain $22x = 46$. The solution to this equation is:

$$x = \frac{46}{22} = \frac{23}{11}$$

Substitute this value in for x in either original equation, and solve for y to obtain $y = \frac{3}{11}$.

2. C: This system is best suited for substitution because the second equation is already solved for the variable y. Plug this equation into the first to obtain:

$$6x - 4\left(\frac{3}{2}x + 9\right) = 14$$

After distributing and collecting like terms, this equation is equivalent to the false statement $-36 = 14$. Because this is never true, there is no solution to the original system of equations.

3. B: The total amount earned in pizza sales is equal to the sum of the price of each pizza times the number of pizzas sold for each size. "At least" is represented by the inequality symbol $\geq$. Therefore, the total sales being at least 800 is represented by the inequality:

$$9m + 10l \geq 800$$

4. 9: For m miles driven, the taxi fare can be represented by the expression $0.75m + 5$. Setting this equal to 11.75 results in the linear equation:

$$0.75m + 5 = 11.75$$

Subtracting 5 and then dividing by 0.75 results in the solution 9.

5. 55: For this example, the revenue function is $R(x) = 2x$, where x represents the number of newspapers sold. In order to make a profit, revenue must be greater than the cost. Therefore, $2x > 0.15x + 100$. Solving this for x results in:

$$1.85x > 100 \text{ or } x > 54.05$$

The answer should be a whole number, as newspapers are being discussed. Therefore, the publisher must sell at least 55 newspapers to make a profit.

6. C: Elimination can be used on this system because of standard form. Multiply the second equation times -2 to obtain:

$$-8x - 16y = -24$$

Adding the equations together results in the identity $0 = 0$, which is always true. Therefore, the two equations represent the same line and any ordered pair on that line is a solution.

7. A: Because 68 people visited the zoo, the sum of adults and kids equals 68 or $a + k = 68$. The total amount paid in admission is equal to the sum of the product of each type of person and the price per person or

$$10a + 8k = 640$$

8. C: To find the equation of the line, the slope needs to be found.

$$m = \frac{64 - 40}{10 - 6} = \frac{24}{4} = 6$$

The slope represents the cost of manufacturing one pair of tennis shoes. Then, plug either ordered pair into the slope-intercept form of a line $y = mx + b$ to find b, the y-intercept. $64 = 6(10) + b$, or $64 = 60 + b$. Therefore, $b = 4$. The equation of the cost function is:

$$C(x) = 6x + 4$$

9. B: To find the unit price, divide the cost by the size in ounces. Thus, $\frac{\$4.50}{32} = \$0.14$ rounded to the nearest cent.

10. 7.98: To solve a proportion, cross-multiply. This step results in the equation $15.96 = 2x$. Dividing both sides by 2 results in the solution $x = 7.98$.

11. 16: There is one teacher for every 8 students. Using this ratio, divide 128 by 8. $128 \div 8 = 16$. Therefore, we would expect there to be 16 teachers at the daycare.

12. 3.25: A proportion can be used to solve this problem. The unknown, x, is the length of the shadow of the house. Therefore, $\frac{16}{4} = \frac{13}{x}$. Cross-multiply to obtain $16x = 52$. Then, divide by 16 to obtain 3.25 feet.

13. 500: Let x be the unknown quantity. Therefore, $.7x = 350$. Divide both sides by .7 to obtain $x = 500$.

14. B: Because the map is smaller than the state, we must scale down. The width of 283 miles would be represented as $\frac{283}{50} = 5\frac{2}{3}$ inches on the map.

15. C: The linear regression model shows a positive correlation between the two variables. When 40 is plugged into the equation for F:

$$C = -0.75(40) + 84.3 = 54.3 \text{ percent calories come from carbohydrates}$$

16. B: The problem can be translated into the equation $.18x = \$215$ where x is the unknown value. To solve for x, divide both sides by .18. Therefore, $\frac{\$215}{.18} = \$1194.44$ rounded to the nearest cent.

17. A: First, convert feet to inches. There are 12 inches in one foot, and therefore, there are $56 \times 12 = 672$ inches in 56 feet. Then, convert 672 inches to centimeters by multiplying times 2.54.

$$672 \times 2.54 = 1,706.88 \; cm$$

Calculator

18. D: The price of Justin's cone plus tip was $1.15x$. The price of Peter's cone was:

$$1.15(x + 1) = 1.15x + 1.15$$

Adding the two together results in $2.3x + 1.15$.

19. 0.491: To solve the inequality, first add $3x$ to both sides. This results in:

$$12x - 4.5 > 1.4$$

Next, add 4.5 to both sides, resulting in $12x > 5.9$. Finally, divide both sides by 12 to obtain $x > 0.491\overline{6}$. The only option that does not satisfy this inequality is 0.491.

20. B: First, reflect the graph over the y-axis. This reflection does not change the location of the y-intercept. Then, by shifting up the line 5 units, the y-intercept is shifted from (0, −5) to (0, 0).

21. C: The options listed are both linear and exponential. Because population is not growing at a constant rate, the model must be exponential. The population is growing, and $P(t) = 2,000(4^t)$ represents growth. Note that $P(0) = 2,000$ and $P(1) = 8,000$.

22. C: The number of houses being built each year is 18, and this value is constant. Therefore, the model is increasing at a constant rate each year. The function that best models this scenario is an increasing linear function.

23. D: For each increase in h, the amount of money M increases by $7.45. Therefore, this amount is the hourly wage of the delivery driver.

24. 40: Because the mean is equal to 30, it is true that:

$$\frac{7 + 19 + x + 54}{4} = 30$$

Therefore, $80 + x = 120$, or $x = 40$.

25. B: 208 is the *y*-intercept of the linear equation. Therefore, when $d = 0$, $H = 208$. If $d = 0$, this means that Chris has not worked any days yet that week. Therefore, the total number of houses that he needs to visit each week is equal to the quantity obtained when plugging $d = 0$ into the equation, resulting in the *y*-intercept.

26. A: The given information can be used to substitute for the variables h and g. This equation then becomes:

$$h = \frac{v^2}{2g} = 11 = \frac{v^2}{2(32)}$$

To solve for the variable v, first multiply by 64. Then raise both sides of the equation to the power $\frac{1}{2}$. Take the square root. The answer is 26.5.

27. C: Standard deviation refers to the spread of the data. The more the data is spread out, the larger the standard deviation. Therefore, because the data in School B is more spread out, the standard deviation is larger.

28. A: The circumference of the circle is 12π and the total circle is 360 degrees, so the arc length for 15 degrees can be found. Set up a proportion to model the relationship: $\frac{15}{360} = \frac{a}{12\pi}$. Solve this proportion by cross-multiplying to find the answer, 0.5π.

29. C: The variable raised to the power $2/3$ is the same as taking the third root of the variable raised to the second power. This can be represented by $\sqrt[3]{x^2} = 64$. For the equation to be solved, the first step is raising both sides to the third power to cancel out the radical. The last step in solving for x is taking the square root of both sides. This results in the number 512.

30. B: This expression required the use of like terms to simplify the problem. The terms in which the x variable has an exponent of 2 are like terms and can be added together. The term with a single x-term and a single constant value does not have like terms and therefore stays the same. Therefore, the correct answer is Choice *B*:

$$-5x^2 + 3x + 4$$

31. A: This equation represents the time it takes for Bill to get to the store by a combination of riding the bus and walking. The store is a distance d away. The first term represents the distance that Bill walked. The second term represents the distance he traveled by bus, $0.5(45 - t)$. The correct answer is Choice *A*:

$$d = 0.04t + 0.5(45 - t)$$

32. C: The equation models how much money you did not account for spending in your checkbook. Since the final amount in the checking account is 75 and you had recorded 87, there must be a missing value of x that you did not record. The correct answer is Choice *C*:

$$87 - x = 75$$

33. B: Use the Pythagorean theorem in this problem to solve for the missing length. The set-up of the equation is:

$$5^2 + 12^2 = c^2$$

Square 5 and 12 and then add both numbers together; the result is 169. Take the square root of 169, which gives the side length of 13.

34. B: The table below shows the fees paid by members and nonmembers.

Number of Visits	1	2	3	4	5	6	7	8
Member's Cost	17	22	27	32	37	42	47	52
Nonmember's Cost	8	16	24	32	40	48	56	64

By the fourth visit, the nonmembers and members are paying the same price. By the fifth visit, the members are paying less and that will continue. If someone is planning to visit the pool more than 4 times, they should pay the member fee.

35. C: Manipulate the first equation by first subtracting 32 from both sides, and then multiply both sides by $\frac{5}{9}$. This allows Darren to convert his temperature from Fahrenheit to Celsius. The correct answer is Choice C:

$$C = \frac{5}{9}(F - 32)$$

36. C: By solving the formula using the two given values to substitute for density and volume, the mass of the substance is found to be 12.48 grams. This was calculated by multiplying the density, 10.4, by the volume, 1.2. The correct answer is Choice C, 12.5 grams.

37. D: The function's value can be found by substituting the value of 2 into the place of the x-value in the equation. The function notation for this becomes:

$$f(2) = 4(2) - 3 = 5$$

The correct answer is Choice D, because $f(2) = 5$.

38. D: Given the type of raise in this job, the equation must be exponential. As the salary increased by a percentage each year, the equation must represent exponential growth. Since the growth was 5%, x needs to be raised to the power of 1.05. The initial salary of 40,000 is found as the coefficient of the term with base 1.05. The correct answer is Choice D:

$$s = 40,000(1.05)^t$$

39. B: To simplify the expression, multiply each term in the first expression by each term in the second expression. The initial combined expression becomes:

$$2x^3 + 2x^2 - 8x - 9x^2 - 3x + 12$$

From this point, collect like terms to form the final expression:

$$2x^3 - 7x^2 - 11x + 12$$

40. C: To solve the equation for the variable x, use the properties of equality. The first step is to move the binomial $(x - 4)$ to the opposite side of the equation by multiplication so that it is no longer in the denominator. Then the equation becomes $6(x - 4) = 12$. Use the distributive property to remove the parenthesis; the equation then becomes $6x - 24 = 12$. To isolate the variable in this equation, add 24 to both sides and then divide by 6. The final answer is $x = 6$.

41. A: The equation models the relationship between height, h, and time, t. The height can be set equal to zero; now the ball is on the ground. This makes the equation:

$$0 = -16t^2 - 16t + 96$$

Solve this equation by factoring; the trinomial then becomes:

$$0 = -16(t + 3)(t - 2)$$

Then, set these two binomials equal to zero. Solve the first factor of $t + 3 = 0$ to yield a time of negative 3 seconds, which is not a real answer. The second factor is $t - 2 = 0$, which gives a time of 2 seconds. Since this is the only positive answer, the time it takes for the ball to hit the ground is 2 seconds. The correct answer is Choice A, 2 seconds.

42. A: The quadratic equation is factored into:

$$y = (x + 7)(x - 2)$$

In order to solve the equation, set the y-value equal to zero. These two binomials give you the two zeros on the graph. The graph opens upward because the coefficient of x^2 is positive. The vertex can be found by the formula $\frac{-b}{2a}$, referring to a and b as the coefficients of x^2 and x, respectively. The correct answer is Choice A, $x = -7, 2$.

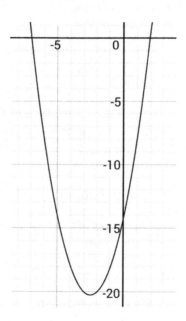

43. B: The expression is simplified by multiplying the binomials by each other. The first step is to use the first, inside, outside, last method, or FOIL method. This results in the expression:

$$(z^2 - 8z + 16)(z - 4)$$

Multiply the polynomials further; each term in the first trinomial must be multiplied by each term in the binomial. Now the like terms can be collected. The final polynomial is Choice B:

$$z^3 - 12z^2 + 48z - 64$$

44. C: Multiplying complex numbers is like multiplying binomials. In accordance with the FOIL method, the constant and the imaginary number in the first binomial must be multiplied by the constant and the imaginary number in the second binomial. The resulting terms are:

$$15 - 9i + 10i - 6i^2$$

Since i is the square root of negative one, $i^2 = -1$. Replacing this value in the final equation and collecting like terms yields the final complex number $21 + i$.

45. B: Finding the sum of two complex numbers is like adding polynomials. The two constants are like terms and can be combined to yield a constant of 7. The two imaginary numbers can be combined to get a value of $3i$. The correct answer is Choice B, $7 + 3i$.

46. B: The volume of a pyramid can be found using the formula $V = \frac{1}{3}lwh$. Given the length, width, and height in the figure, the equation is:

$$V = \frac{1}{3}(4)3(4)$$

Multiplying these values yields a final result of 16 cubic centimeters. The correct answer is Choice B, 16 cm³.

47. B: The x-coordinate of point B represents $\cos(\frac{3\pi}{4})$. Since $\frac{\pi}{2}$ lies on the positive y-axis and π lies on the negative x-axis, then $\frac{3\pi}{4}$ lies in between those two axes, in the second quadrant of the graph. The cosine refers to the adjacent portion of the angle, which is the x-coordinate. The correct answer is Choice B, in which the x-coordinate is B.

48. A: To find the function value, replace the input variable with -3. The equation becomes:

$$f(-3) = (-3)^3 + 3(-3)$$

Solving this equation gives the function value of -36.

Free Exam Tips Videos/DVD

We have created a set of videos to better prepare you for your exam. We would like to give you access to these **videos** to show you our appreciation for choosing Exampedia. **They cover proven strategies that will teach you how to prepare for your exam and feel confident on test day.**

To receive your free videos, email us your thoughts, good or bad, about this book. Your feedback will help us improve our guides and better serve customers in the future.

Here are the steps:

    1. Email **freevideos@exampedia.org**

    2. Put **"Exam Tips"** in the subject line

Add the following information in the body of the email:

    3. **Book Title:** The title of this book.

    4. **Rating on a Scale of 1–5:** With 5 being the best, tell us what you would rate this book.

    5. **Feedback:** Give us some details about what you liked or didn't like.

Thanks again!

Made in the USA
Las Vegas, NV
23 July 2021